Ashwini Sri Hari
Siva Prasad Bitragunta
Sankar Ganesh Palani

Assessment of Titanium dioxide nanoparticle toxicity in earthworms

AF154062

Ashwini Sri Hari
Siva Prasad Bitragunta
Sankar Ganesh Palani

Assessment of Titanium dioxide nanoparticle toxicity in earthworms

Ecotoxicity of Titanium dioxide nanoparticles

LAP LAMBERT Academic Publishing

Impressum / Imprint

Bibliografische Information der Deutschen Nationalbibliothek: Die Deutsche Nationalbibliothek verzeichnet diese Publikation in der Deutschen Nationalbibliografie; detaillierte bibliografische Daten sind im Internet über http://dnb.d-nb.de abrufbar.

Alle in diesem Buch genannten Marken und Produktnamen unterliegen warenzeichen-, marken- oder patentrechtlichem Schutz bzw. sind Warenzeichen oder eingetragene Warenzeichen der jeweiligen Inhaber. Die Wiedergabe von Marken, Produktnamen, Gebrauchsnamen, Handelsnamen, Warenbezeichnungen u.s.w. in diesem Werk berechtigt auch ohne besondere Kennzeichnung nicht zu der Annahme, dass solche Namen im Sinne der Warenzeichen- und Markenschutzgesetzgebung als frei zu betrachten wären und daher von jedermann benutzt werden dürften.

Bibliographic information published by the Deutsche Nationalbibliothek: The Deutsche Nationalbibliothek lists this publication in the Deutsche Nationalbibliografie; detailed bibliographic data are available in the Internet at http://dnb.d-nb.de.

Any brand names and product names mentioned in this book are subject to trademark, brand or patent protection and are trademarks or registered trademarks of their respective holders. The use of brand names, product names, common names, trade names, product descriptions etc. even without a particular marking in this works is in no way to be construed to mean that such names may be regarded as unrestricted in respect of trademark and brand protection legislation and could thus be used by anyone.

Coverbild / Cover image: www.ingimage.com

Verlag / Publisher:
LAP LAMBERT Academic Publishing
ist ein Imprint der / is a trademark of
OmniScriptum GmbH & Co. KG
Heinrich-Böcking-Str. 6-8, 66121 Saarbrücken, Deutschland / Germany
Email: info@lap-publishing.com

Herstellung: siehe letzte Seite /
Printed at: see last page
ISBN: 978-3-659-55283-0

DEDICATION

In loving memory of Mr.A.T.Sridharan and Mr.T.Ponnuswamy who have been exemplary examples of hard work and honesty and whose legacy I intend to carry forward....

ACKNOWLEDGEMENT

I would like to convey my heartfelt gratitude to my Project Investigator and mentor, Dr.P.Sankar Ganesh for having accepted me as his thesis student and providing unconditional support and guidance with regard to all aspects of my research and career.

I would like convey my heartfelt gratitude to my Doctoral candidate mentor, Mr. B. Siva Prasad for introducing me to the upcoming and exciting field of Nanotoxicology and for the unconditional support and guidance extended with regard to all aspects of my work and career.

I would also like to thank the Heads of the Department of Biological Sciences and Pharmacy, Dr. Ramakrishna Vadrevu and Dr. Vamsi Krishna respectively, for providing the facilities required to carry out my research work.

I would like to extend my heartfelt gratitude to my parents and friends for giving me constant support and encouragement that aided in the successful completion of my thesis.

Table of Contents

List of symbols and abbreviations

µL – microliter

Ag – Silver

Al(OH)$_3$ - Aluminium hydroxide

AMT – Advanced Microscopy Techniques

Al$_2$O$_3$ – Aluminium oxide

Au - Gold

BET – Brauner Emmett Teller

BSA – Bovine Serum Albumin

BSE – Back scattered electrons

CAT – Catalase

CeO$_2$ – Cerium (IV) oxide

CuO – Copper (II) oxide

CuSO$_4$ – Copper (II) sulphate

d(H) - hydrodynamic diameter

d - Day

D - translational diffusion coefficient

DLS - Dynamic Light Scattering technique

DNA – Deoxyribonucleic acid

EDTA – Ethylene diamine tetra acetic acid

ENMs – Engineered nanomaterials

ENPs – Engineered nanoparticles

EPA – Environmental Protection Agency

Fe$_2$O$_3$ – Iron (III) oxide

g/Kg – gram per kilogram

GR – Glutathione reductase

GSH - glutathione

GSSG - glutathione disulfide

h - Hour

H$_2$O$_2$ - Hydrogen peroxide

HCl – Hydrochloric acid

ICP-MS – Inductively coupled plasma-mass spectrophotometry

k - Boltzmann's constant

KCl – Potassium chloride

kV – Kilovolt

LC$_{50}$ – Lethal concentration 50

m^2/g – meter square per gram

MCDA - Multi-Criterion Decision Analysis

MDA – Malondialdehyde

mg/cm^2 – milligram per centimeter square

mg/Kg – milligram per kilogram

mg/L – milligram per litre

mg/mL – milligram per milliliter

mM – millimolar

mW – milliwatt

NA – Not Applicable

NADPH - Nicotinamide Adenine Dinucleotide Phosphate reduced form

nm – nanometer

NP – Nanoparticle

^{o}C – Degree Celsius

OECD – Organization for Economic Co-operation and Development

PDI – Polydispersity index

PDMS - Poly dimethyl siloxane

POD – Peroxidase

RCEP – Regional Comprehensive Economic Partnership

ROS – Reactive Oxygen Species

SEM - Scanning Electron Microscopy

SiO$_2$ – Silicon dioxide

SOD – Superoxide dismutase

SSA – Specific Surface Area

T - Absolute temperature

TBARS – Thiobarbituric acid reacting substances

TEM – Transmission Electron Microscopy

THF – Tetrahydrofuran

TiO$_2$ – Titanium dioxide

UNEP – United Nations Environment Programme

UV – Ultraviolet

ZnO – Zinc oxide

ZrO$_2$ – Zirconium dioxide

η – viscosity

Chapter 1: Introduction

In the 21st century, science of nanotechnology has gained irrefutable importance due to its versatility and applicability in a variety of domains such as healthcare, pharmaceuticals (drug delivery), cosmetics (sunscreens), energy (solar cells, fuel cells), textiles, agriculture, electronics (sensors), communication, computing (information storage) and environmental remediation to name a few. Nano-based products are replacing their bulk counterparts in conventional goods at an accelerating pace. Engineered nanomaterials (ENMs) are conventional materials deliberately engineered to be nanostructured and are being used in a number of sectors as they exhibit unique physical and chemical properties that are attributed to their small size. Examples of ENMs include carbon nanotubes, metal and metal oxide based nanomaterials (Ag, Au, Al_2O_3, CeO_2, CuO, SiO_2, TiO_2, ZrO_2 etc.) dendrimers, and nanocomposites. At nano level, effects of surface and quantum characteristics increase thereby becoming evident and giving rise to enhanced chemical reactivity. Hence the optical, electrical, mechanical and magnetic properties of engineered nanomaterials are significantly different from bulk materials. Enhanced chemical reactivity of engineered nanomaterials makes it difficult to predict the nature of their interactions, behavior, fate, transport etc., in different matrices ultimately giving rise to uncertainties in their risk assessment. The emerging field of nanotoxicology is an effort to assess the potential risks posed by nanomaterials. The central question, therefore, is whether the unknown risks of ENMs, in particular their social and environmental impacts outweigh their established benefits for society (Vicki., 2003). Fate of ENMs after their intended use is unknown and not well documented. ENMs will ultimately find their way into environment thereby adversely affecting different ecosystems and ultimately the human health. Lack of appropriate technical data on ecotoxicity of ENMs has been the barrier for establishing moratorium on nanotechnology research and regulation of nanomaterials. For the development of reliable and robust risk assessment framework and regulatory guidelines for ENMs, many essential parameters such as accurate physicochemical characterization, appropriate dose metric, agglomeration characteristics in solvents, solubility characteristics etc., are to be defined and determined. These parameters will be predictive about the interaction of ENMs with physiological and environmental matrices. This would undoubtedly aid in understanding realistic exposure scenarios of organisms representative of different ecosystems and their concordant responses to a diverse array of ENMs. At the organism level, however, sensitive endpoints are to be identified that would be reliable indicators of toxicity of ENMs. Some examples of the endpoints that have already been explored in different model organisms are activity profiles of antioxidant enzymes, lipid peroxidation, genotoxicity (DNA damage), frequency of apoptosis etc. In addition to this, data from *in vitro* and characterization studies would provide a comprehensive outlook on the ecotoxicity of ENMs and would lay the foundation for reliable risk assessment strategies.

Rutile titanium dioxide nanoparticles are one of the three forms of TiO_2. At the nano level, TiO_2 exhibits enhanced surface and quantum effects due to tight control of particle size (Robichaud *et al.,* 2009). These effects also contribute to properties such as high stability, high transparency to visible light, high UV absorption, anticorrosiveness and photocatalytic effect due to which it is widely used in cosmetics, pharmaceutics and paint industries. The massive applications of TiO_2 nanoparticles in different fields will inherently result in their release into the environment and thereby exposure to organisms (Kaegi *et al.,* 2008). Based on the potential environmental risk, it is important to quantify the environmental fate and transport behavior of TiO_2 nanoparticles (Si Li *et al.,* 2011). Recent literature reviews have demonstrated the lack of knowledge in this field, with especially little data being available on effects of ecotoxicity of nano TiO_2 on soil invertebrates (Handy *et al.,* 2008; Klaine *et al.,* 2008; Kahru and Dubourguier., 2010). Moreover well established and validated protocols for chemical risk assessment are unavailable to carry out reliable and accurate ecotoxicity assessment of TiO_2 nanoparticles in the terrestrial ecosystem. Earthworm (*Eisenia fetida*) is one of the extensively studied indicator organisms of environmental pollution in the terrestrial ecosystem and is exposed to anthropogenic compounds through both its external and internal surfaces (Sussana *et al.*, 2012).

Over the last few decades, a large number of useful biomarkers applicable both under laboratory and field conditions have been developed in earthworms (Scott *et al.* 2000; Spurgeon *et al.,* 2002). Among the biomarkers developed are three oxidative stress related enzymes catalase, glutathione reductase and superoxide dismutase. These antioxidant enzymes could serve as potential biomarkers for assessing and monitoring TiO_2 nanoparticle ecotoxicity in the terrestrial ecosystem. The activity profile of these enzymes in response to oxidative stress that would be generated after uptake of these nanoparticles would be correlated to nanoparticle concentration and physicochemical characteristics. This correlation would provide a deeper perspective about the behavior of TiO_2 rutile nanoparticles and the consequent response of the chosen biomarkers in the model organism. Based on appropriateness of response and sensitivity of the biomarkers, their feasibility in being used for TiO_2 nanoparticle ecotoxicity assessment would be determined.

In the current study, ecotoxicity of titanium dioxide (TiO_2) nanoparticles on earthworm species *Eisenia fetida* has been studied taking into account characterization of nanoparticles and response of the organism to antioxidant enzymes superoxide dismutase (SOD), glutathione reductase (GR), catalase (CAT) and lipid peroxidation.

9

Chapter 2: Review of Literature

Ecotoxicological evaluation of TiO₂ nanoparticles-State of the art

2.1 The Nanotechnology era

In December 1959, the concept of miniaturization and consequent manipulation of matter at the atomic scale was conceived by Dr. Richard Feynman, an American physicist, thereby fuelling the dawn of the era of nanotechnology. Nevertheless it was not until several decades later that nano based consumer products were introduced into the market. After their introduction, the number of nano-based goods has increased in an almost exponential manner with a total of 1628 different products as on October 2013.

(Project on Emerging Nanotechnologies (2014), http://www.nanotechproject.org/cpi)

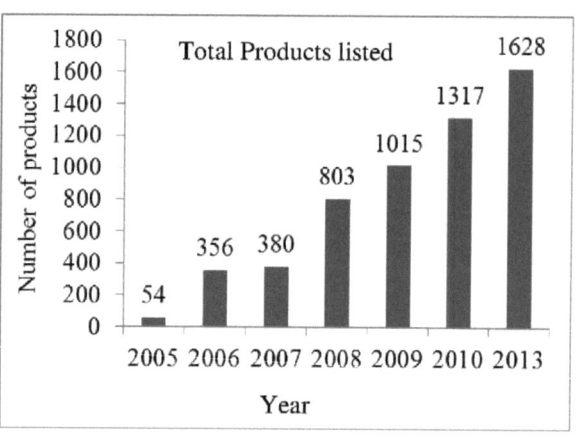

Figure.1 Nano based products in the market

(Source:http://www.nanotechproject.org/cpi/site/assets/files/3551/chart_1.png)

Nanoscale particles, however, already exist in the biosphere as a result of natural processes and accidental releases from anthropogenic processes. But they are different from engineered nanoparticles (ENPs) which are deliberately synthesized from bulk substances to be nanosized. ENPs possess specific and often enhanced physicochemical properties such as high specific surface area, high surface energy and increased reactivity in comparison with their bulk counterparts. It is these improved features of ENPs that manufacturers seek to exploit thereby increasing the global sales of nano-based products. These ENPs along with numerous bulk substances are the building blocks of engineered nanomaterials (ENMs) (Oberdörster, 2005). ENMs are intentionally designed and produced to have a size typically between 1-100nm in 1, 2 or 3 dimensions and to exhibit specific properties and uniform structures (Source : www.epa.gov, 2007). A variety of ENMs are being employed for a number of applications in sectors such as healthcare, pharmaceuticals, cosmetics, energy, environmental remediation, textiles, agriculture, electronics, communication and computing to name a few because of their improved surface and quantum effects. The U.S Environmental Protection Agency (U.S EPA) classifies ENMs into 4 types: carbon based, metal and metal oxide based, dendrimers and quantum dots. Amongst these, carbon and metal and metal oxide based nanomaterials are termed generation I or passive nanostructures which are nanomaterials that do not respond to changes in the environment but add functionality to a system (http://www.nano.monash.edu.au/case-studies/terry-turney.html). Some examples include nanoparticles such as (i) TiO_2 in sunscreens, anti-aging creams, food additives, wrinkle-free textiles, (ii) iron oxide nanoparticles as additives in construction materials and industrial coatings, (iii) zinc oxide calcium alginate nanofilms as food preservative, and (iv) superhydrophobic carbon nanotubes for separating oil and water. These materials have steady or quasi-steady structures and functions; such as mechanical behavior and chemical reactivity (http://nice.asu.edu/mechanism/passive nanostructure).Amongst the generation I nanomaterials, metal and metal oxide based nanomaterials have gained significant importance due to their diversity of applications in commercial products thereby augmenting their contribution to the global economy.

11

2.2 Significance of metal and metal oxide nanoparticles

According to recent statistics from the 'The Project on Emerging Nanotechnologies', three out of four products released per week may contain metal oxide nanomaterials. They have surpassed even carbon based nanomaterials in the market. Metal and metal oxide based nanomaterials have gained significant importance due to their versatility in application. Metal oxide nanomaterials in particular are growing assets in technological industries as they can adopt a vast number of structural geometries with an electronic structure that can exhibit metallic, semiconductor or insulator character and hence are used in the fabrication of sensors, microelectronic circuits, as coatings to protect surfaces against corrosion and as catalysts (Marcos Fernández-Garcia, José A. Rodriguez., 2007). Apart from this, metal oxide based nanomaterials have been introduced into a variety of consumer products such as cosmetics and sunscreens (TiO_2, Fe_2O_3 and ZnO) (Nowack and Bucheli., 2007), fillers in dental fillings (SiO_2) (Balamurugan et al., 2006), in catalysis (TiO_2) (Aitken et al., 2006), and as fuel additives (CeO_2) (Laosiripojana et al., 2005) (Yon Ju-Nam and Jamie Lead., 2008).

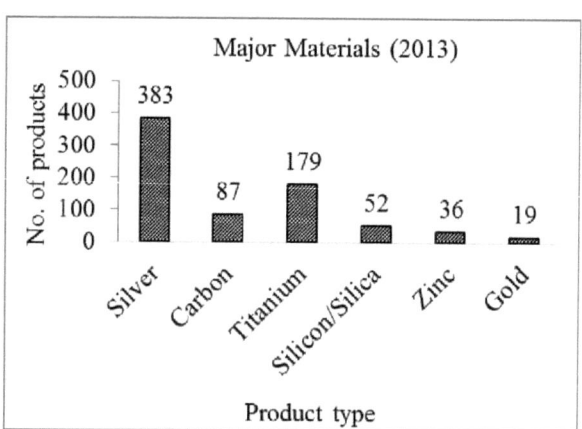

Figure.2 Distribution statistics of various ENMs in the year 2013
(http://www.nanotechproject.org/cpi/about/analysis/)

With the increased presence of metal and metal oxide nanomaterials in commercial products, exposure to environment and humans is inevitable. As the incorporation of metal and metal oxide nanoparticles in consumer products increases, their release into different environmental sinks also increases. Therefore it necessary to develop strategies to detect, quantify and thereby assess the impact of such nanoparticles in different environmental compartments.

2.3 Fate of engineered nanoparticles in the environment

Environment (including aquatic and terrestrial ecosystems) is the ultimate sink for the ENMs after their intended use. Some common routes of entry of ENMs into the environment would include effluents from industries, sewage, agricultural runoffs (fertilizers containing ENMs), environmental remediation and land reclamation (using ENMs) etc. Evaluation of potential hazards related to exposure to various ENMs has thus become an emerging field of toxicology and environmental health (David. B. Warheit., 2007). Current and future research efforts in this field of toxicology (nanotoxicology) are largely aimed at establishing (eco) toxicological and exposure data of ENMs (Khara. D. Grieger *et al.,* 2009). The existing data from pulmonary toxicity studies of ultrafine particles and the risk assessment framework for chemicals and bulk materials will help form the basis for studying ENMs. However the main challenge would be to address the fate, behavior and transport of ENMs in various environmental matrices (soil & water) which in turn are interdependent on the physicochemical properties of the nanoparticles that govern their mobility, reactivity and toxicity. These uncertainties together with challenges involved in detection and quantification of ENMs have to be addressed in order to assess their ecotoxicity and develop a risk assessment framework for ENMs.

2.4 Broad overview on challenges faced in ENM ecotoxicity assessment

Experimental data about acute and chronic toxicity and environmental fate of engineered nanomaterials score lower by far: tested materials are less than 20% in this respect. Ecotoxicity assessment of ENMs relies heavily on accurate physical and chemical characterization of these materials. This concern has been addressed by many authors (Oberdorster *et al.,* 2005; Powers *et al.,* 2006 & 2007; Klaine *et al.* 2008; Warheit., 2008; Domingos *et al.,*2009) and a list of essential characterization parameters has been developed. The characterization challenges include accurate determination of size, surface properties, aggregation characteristics & mechanisms, solubility characteristics, variability in ENM preparation, behavior & interaction of ENMs in various matrices (Michala. E. Pettitt and Jamie. R. Lead., 2013). Apart from characterization concerns, toxicity studies in terms of exposure, dose, hazard and risk are to be studied for engineered nanoparticles. To date, however, nanoparticle toxicity studies have mainly focused on *in vitro* examinations due to the ease in execution, control and interpretation of the experiments compared to *in vivo* tests. Therefore, more consideration must be placed on improving *in vivo* experimental procedures of nanoparticle toxicity assays in order to gain reliable data (Kong *et al.*, 2011). Methods of toxicity and ecotoxicity assessment of nanomaterials should essentially comprise a list of short and long term effects on species that inhabit the aquatic and terrestrial habitats (Gourmelon and Ahtiainen., 2007). Firstly, acute toxicity studies in the sentinel organisms should be performed, which should be followed by studies that focus on morphological, ethological and biochemical changes. This would provide a comprehensive outlook on the ecotoxicity of nanoparticles in a given ecosystem. All these intricate requirements that are specific to ENMs make it difficult to adopt existing methodologies such as OECD guidelines for the risk assessment of ENMs and hence require a comprehensive and strategic framework for the assessment of ecotoxicity of ENMs.

There is an unprecedented need for establishing such a comprehensive and robust risk assessment framework especially for metal oxide based nanomaterials that are rapidly proliferating in the consumer markets. Nano titanium dioxide is one of the most widely used metal oxide nanoparticles till date (Robichaud *et al.*, 2009). An estimated 2000 metric tons of nano TiO_2 worth $70 million were produced in the year 2005 (EPA, U. External Review Draft - Nanomaterial Case studies). In the year 2010, the production had increased to 5000 metric tons and is expected to increase by leaps and bounds by the year 2025 (Landsiedel *et al.*, 2010). These nanoparticles have been incorporated into a number of branded consumer products such as Neutrogena sensitive sunblock cream, M & M's chocolate candy, Colgate toothpaste, Mentos mint, Nestle coffee creamer etc. The study conducted by Alex Weir *et al.*, (2012) highlights the presence of nano TiO_2 in a variety of food and personal care products.

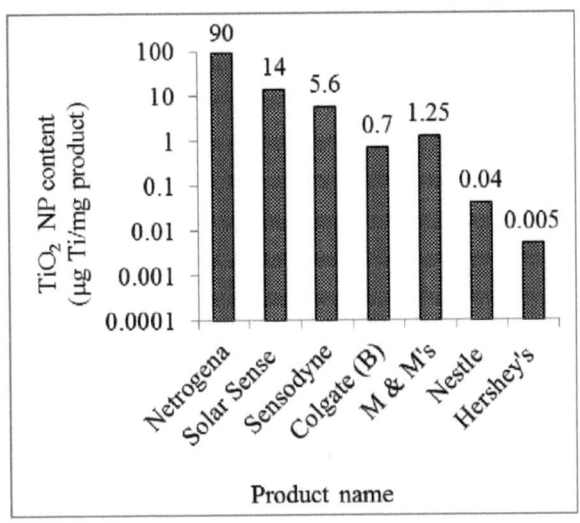

Figure.3 Examples of some personal care and food products containing nano TiO_2

(Alex Weir *et al.*, 2012)

With such widespread usage of TiO_2 nanoparticles, the occurrence of nano TiO_2 in the environment and interaction with various ecosystems is inevitable. It is therefore highly necessary to detect, quantify and assess the toxicological impact of these nanoparticles in different environmental matrices.

2.5 Need for ecotoxicity assessment of TiO$_2$ nanoparticles

Titanium dioxide, also known as titanium (IV) oxide or titania is a natural oxide of titanium. It is a naturally occurring mineral and exists in 3 crystalline forms: anatase, rutile and brookite and an amorphous form (Reyes-Coronado et al., 2008). TiO$_2$ is a transition metal oxide and is known as the principal white pigment of commerce. The properties that make it predominant in the market are high refractive index, lack of colour and chemical inertness. Nano-TiO$_2$ was one of the first nanomaterials to be adopted by a wide variety of industries due to its improved physical and chemical properties at the nanoscale. It has been estimated that the commercial production of nano-TiO$_2$ would be close to 10,000 metric tonnes per year between 2011 and 2014 (UNEP, 2007). Amongst the metal oxide nanoparticles used in commercial products, titanium dioxide nanoparticles (TiO$_2$ NPs) are by far the most exploited according to October 2013 statistics from 'The Project on Emerging Nanotechnologies'.

The rutile phase of TiO$_2$ has the highest refractive index and relative scattering power amongst the three forms (http://www.rsc.org/). Due to this it is used to provide opacity to paints, inks, toothpaste etc. It is used as a pigment and thickener in cosmetic products. Another useful property is the capacity to absorb ultraviolet light (UV) light (even potentially damaging UV light) and hence it is used in plastics, sunscreens and other applications. The anatase phase on the other hand has comparatively less refractive index and light scattering capacity but is well known for its photocatalytic property and is hence used in catalysis and photocatalysis applications (Mueller and Nowack, 2008). On a global scale nano TiO$_2$ (rutile form) has much wider applications especially in consumer products in comparison with the anatase form. Due to all the above mentioned reasons, titanium dioxide nanoparticle release into the environment and potential hazards related to the consequent exposure are inevitable (Hall et al., 2009). This is one of the major reasons for TiO$_2$

nanoparticles being the most extensively studied metal oxide nanoparticles in terms of ecotoxicity assessment (Cattaneo *et al.*, 2009; Kahru and Dubourguier, 2010).

2.6 Significance of terrestrial invertebrates in TiO$_2$ nanoparticle ecotoxicity assessment

The terrestrial ecosystem is one of the major sinks for ENMs. However little data are available with regard to environmental relevant concentrations for TiO$_2$ nanoparticles in the terrestrial ecosystem. The major routes of entry of TiO$_2$ nanoparticles into this ecosystem would be application of sewage sludge on land, effluent from waste water treatment plants, leachates and surface runoffs from exterior façade paints etc. (Anja Menard *et al.*, 2011). The concentration of metallic Ti in surface runoffs was estimated by Kaeigi *et al.*, (2008) to be 600µg/L. In raw sewage the concentration of Ti was found to be as high as 100-3000µg/L (Kiser *et al.*, 2009) and 5-15µg/L in effluents from wastewater treatment plants. The tendency of nanoparticles to aggregate and sediment after being introduced into water has been reported by Boxall *et al.*, (2007). This could facilitate the interaction between nanoparticles and soil dwelling organisms. The exposure of these organisms and hence bioavailability of nanoparticles would depend on several factors such as concentration, size, surface characteristics, state of aggregation and agglomeration as well as adsorption to various environmental media such as sediments, biofilms etc. (Karl Fent., 2010). Till date there is only little data available on ecotoxicity of TiO$_2$ nanoparticles in the terrestrial ecosystem that addresses the influence of some of the factors mentioned on the interaction of these particles with terrestrial biota. This is mainly attributed to the lack of separation and analytical methods to quantify environmental concentrations and to characterize the physical status, the adsorption/desorption, the precipitation and the dissolution of engineered nanoparticles after being released into the soil. Future emission rates of such engineered nanoparticles are still unknown and scarce ecotoxicological data exists, compromising the ability to predict their impact on terrestrial communities (Mueller and Nowack., 2008, Roh *et al.,* 2010). Only fewer have focused on the mobility and transport of ENPs into the soil matrix (Doshi *et al.,* 2008). Gourmelon and Ahtiainen., (2007) have emphasized that the ecotoxicity assessment of nanomaterials should essentially comprise a list of short and long term effects on species that inhabit the aquatic and terrestrial habitats. Hence the preliminary studies on exposure assessment of nanomaterials should ideally begin with invertebrate models of any ecosystem followed by organisms representative of higher trophic levels in the food chain. This would help in understanding the effect of physicochemical characteristics of the nanoparticles on simple organisms and allow parameters like bioaccumulation to be studied as we move to different trophic levels in the food chain. In this regard, studies focusing on ecotoxicity of nano TiO_2 on terrestrial invertebrates are limited compared to aquatic invertebrates. Till date few studies only have been documented with regard to ecotoxicity of TiO_2 nanoparticles on terrestrial invertebrates such as oligochaetes (earthworms), nematodes (*Caenorhabditis elegans*) and isopods (*Porcellio scaber)*. Amongst these invertebrate organisms, earthworms have long been considered for ecotoxicological testing of industrial chemicals and pesticides by global organizations such as Organization for Economic Cooperation and Development (OECD) and Food and Agricultural organization (FAO) and also possess highly differentiated organs, tissues and an immune system almost comparable to humans despite belonging to the lower class of invertebrates and hence would serve as an appropriate indicator organism of ecotoxicity of TiO_2 nanoparticles in the terrestrial ecosystem.

2.7 Significance of earthworm as an invertebrate model organism for TiO_2 ecotoxicity assessment

The use of earthworms for toxicity testing is highly recommended by many toxicologists and is considered as much preferred indicator organisms for monitoring pollution in the environment (Venkateswara Rao and Kavitha., 2004; Reinecke and Reinecke., 2007; Zhou *et al.,* 2007). Earthworms are key species within soils and are exposed to anthropogenic compounds through both their external and internal surfaces. For those reasons, these invertebrates are useful indicators of soil quality and are widely used as model organisms in terrestrial ecotoxicology (Sussana *et al.*, 2012). Over the last few decades, a large number of useful biomarkers applicable under both laboratory and field conditions have been developed in earthworms (Scott *et al.*, 2000; Spurgeon *et al.*, 2002). *Eisenia fetida* and *Eisenia andrei* are commonly used for standard toxicity tests (OECD 207, 222 guidelines; ISO 17512-1, soil quality) and ecotoxicological studies. Recently, a battery of biomarkers suitable for proving pollutant-induced physiological changes at different levels of functional complexity has been developed on these earthworm species. Since little information is available on the toxicity produced by engineered nanoparticles, simple *in vitro* toxicity models are need of the hour. Earthworms as model organisms, to study the ecotoxicological effects as per the OECD 207 guidelines, have been widely used in many studies. But such studies scarcely have focused on engineered nanoparticles as toxicants.

Studies focusing on ecotoxicity of TiO_2 nanoparticles with earthworm as the test organism are scarce. In the study by Hu *et al.*,(2010) the earthworm species *Eisenia fetida* was exposed to TiO_2 rutile nanoparticles having an average diameter of 10 - 20nm in artificial soil test (OECD 207 guideline) for a duration of 7 days. The concentrations of TiO_2 nanoparticles used were 0.1, 0.5, 1 and 5mg/Kg in artificial soil containing distilled water. The endpoints such as bioaccumulation, activity of enzymes such as SOD, CAT, cellulase, lipid peroxidation and genotoxicity were estimated. It was observed that TiO_2 nanoparticles were indeed toxic to the worms when the concentration exceeded 1g/Kg. Bioaccumulation was observed beyond 5g/Kg. The activity of cellulase decreased with increase in concentration of TiO_2 nanoparticles. The authors had also studied the activity of antioxidant enzymes, lipid peroxidation and DNA damage as oxidative stress through ROS production was thought to be the main mechanism of toxicity of nanoparticles (Kohen and Nyska.,

2002). Activity of CAT was highest at 1g/Kg but a decrease in activity was observed at 0.5g/Kg. Activity of SOD was highest at 0.5g/Kg after which a gradual decrease in activity with increase in concentration was observed. This indicates that there is no clear dose-response relationship between enzyme activity and concentration of nanoparticles. However at higher concentrations (1g/Kg) the activity of SOD and CAT decreased slightly as the antioxidant defense system was overwhelmed (Nel *et al.,*2006). As a result, lipid peroxidation indicated by increase in MDA content for the first three concentrations and DNA damage at 1g/Kg and 5g/Kg were observed. The authors conclude that TiO_2 nanoparticles could induce harmful effects and also bioaccumulate within the worm.

In the study done by Canas *et al.,* (2011), acute and reproductive toxicity of TiO_2 nanoparticles (anatase) to *Eisenia fetida* were studied for duration of 14 days by filter paper contact test, sand acute test and artificial soil test. The concentrations chosen for study were 0.1, 1, 10, 100, 1000, 5000 and 10,000 mg/L. The nanoparticles in suspension were characterized by Scanning Electron Microscopy (SEM) and Dynamic Light Scattering (DLS) and the aggregation of particles with increase in concentration was evident in the concentration range from 1-100mg/L. Earthworms survived at all concentrations of TiO_2 nanoparticles on filter paper. In sand, significant acute toxicity of TiO_2 nanoparticles to the worms was not observed. In artificial soil test, a decrease in cocoon production with increase in concentration was evident, however a significant dose-response relationship could not be observed. The possible mechanisms of toxicity of TiO_2 nanoparticles indicated aggregation of particles that could lead to dissimilar biological activity and the generation of hydroxyl radicals by TiO_2 nanoparticles either in the presence or absence of UV light.

Lapied *et al.,* (2011) studied the bioaccumulation of an aged TiO_2 nanocomposite coated with $Al(OH)_3$ and PDMS (Poly dimethyl siloxane) in the model organism *Lumbricus terrestris* and its effect on the frequency of apoptosis in the organism. The study was carried out in three different exposure media with the following concentrations of TiO_2 nanoparticles: Water (1, 10, 100mg/L); dry food (10 & 100mg/Kg); soil (15mg/Kg). The duration of exposure was 7 days (water) and 2-8 weeks (soil). Characterization studies were not carried out. The apoptotic frequency was highest at 100mg/L in water as exposure media. In soil at 15mg/Kg an increase in apoptosis was observed in cuticle tissue. The authors suggest that there is no significant difference in the bioavailability of the compound in water and soil. They also highlighted that apoptosis is tissue specific and it could be a better and more sensitive endpoint for ecotoxicity assessment of nanoparticles. No bioaccumulation of TiO_2 nanocomposite was reported. The authors conclude by suggesting that coated TiO_2 nanocomposites have reduced propensity of ROS generation that would ultimately lead to oxidative stress which is one of the main mechanisms of toxicity of nanoparticles.

In the study done by McShane *et al.,* (2012), two test organisms *Eisenia fetida* and *Eisenia andrei* were exposed to TiO_2 nanoparticles of different sizes and composition (5nm-100% anatase; 10nm-100% anatase & 21nm-80% anatase and 20% rutile). Field test and artificial soil test were carried out for a concentration range of 200-10,000mg/Kg and endpoints such as mortality, reproduction, juvenile growth and avoidance behavior were assessed. Characterization of the particles was done by Transmission Electron Microscopy (TEM), Dynamic Light Scattering (DLS) and Brunauer Emmett and Teller (BET) method. The authors report that survival of earthworm adults as well as their offsprings was not affected by TiO_2 nanoparticles even at concentrations as high as 10,000mg/Kg. The authors recognize the effects of constituents of a complex matrix such as soil on the characteristics and behavior of the particles. Earthworms avoided the soil amended with nano TiO_2 at concentrations from 1000mg/Kg and higher. No avoidance behavior was observed in soil amended with micro-sized TiO_2 (control). The authors thereby emphasize the role of particle size and specific surface area in the avoidance behavior of the worms. The authors also do not rule out the possibility of ROS generation as an indirect cause of avoidance behavior. The authors also suggest that soil chemistry and adsorptive properties of the nanoparticles could have significant influence on behavior of the worms.

On the basis of the studies discussed above, the following insights are evident.

- TiO_2 nanoparticles can undergo agglomeration in suspension thereby influencing particle size
- TiO_2 nanoparticles are capable of inducing oxidative stress in the worms
- There is no clear dose-response relation between activity of antioxidant enzymes and concentration of TiO_2 nanoparticles
- Lipid peroxidation and genotoxicity are additional indicators of induced oxidative stress
- The exposure medium can influence the bioavailability and physicochemical characteristics of TiO_2 nanoparticles

2.8 Current gaps in research specific to TiO_2 nanoparticle ecotoxicity studies on earthworms

From the studies that have been documented on the ecotoxicity of TiO_2 nanoparticles on earthworms, research gaps are evident in the following domains that need to be addressed independently and in combination with each other in future toxicity studies.

- Characterization of nanoparticles

- Organism characteristics

- Methodology employed for toxicity assessment

The physicochemical characteristics of TiO$_2$ nanoparticles such as size, surface area, surface charge, agglomeration, and dissolution should be ascertained before and after introduction of TiO$_2$ nanoparticles into the exposure medium. The influence of various medium components, pH, ionic strength, nature of electrolytes, etc., on the physicochemical characteristics of TiO$_2$ nanoparticles should be established in order to understand various phenomena such as bioavailability of the nanoparticles, uptake of nanoparticles by the organism and subsequent dose-response characteristics exhibited by the organism. These phenomena are also influenced by TiO$_2$ nanoparticle type, the presence/absence of coating on the nanoparticle, type of organism used, method of application of TiO$_2$ nanoparticles to the exposure medium and the type of exposure medium. The existing studies on ecotoxicity of TiO$_2$ nanoparticles to earthworms have addressed the influence of only some of the parameters (stated above) on the toxicity of these nanoparticles to earthworms.

Nanoparticle characteristics:

According to Stone et al., (2010), six parameters have been prioritized for ecotoxicity assessment of nanoparticles. They are size, dissolution, surface area, surface charge, and surface chemistry. Table I (Supplementary data) indicates the type of characterization undertaken in each of the studies performed. Out of the 4 studies, only one (Mc.Shane et al., 2012) had reported near to complete characterization of the TiO$_2$ nanoparticles in terms of particle size, agglomeration and specific surface area. In most of the studies soil has been used as the exposure medium. However the behavior of nanoparticles in such complex matrices could not be determined.

Moreover matrix parameters like pH, ionic strength, nature of electrolytes (Navarro et al., 2008; Sharma, 2009) and the presence of organic acids (Pettibone et al., 2008; Domingos et al., 2009) should be taken into account as they could control agglomeration of the particles. Also in the study that employed filter paper test (Canas et al., 2011), the interaction of TiO_2 nanoparticles with constituents of the filter paper was not taken into account. All these parameters would influence particle size and uptake by the worms. Two other characteristics of TiO_2 nanoparticles such as type (anatase/rutile) and the presence or absence of external coating on the nanoparticle would also dictate the interaction of the nanoparticles with the exposure medium, earthworms and the degree of oxidative stress induced and hence should be taken into consideration during ecotoxicity assessment of TiO_2 nanoparticles.

Organism characteristics:

Three earthworm species (*Eisenia fetida, Eisenia andrei and Lumbricus terrestris*) have been considered for ecotoxicity assessment of TiO_2 nanoparticles till date. Though the organisms belong to the same class, species level difference would have some impact on their behavior and response to TiO_2 nanoparticles. Hence this could also be attributed as one of the reasons for dose-response variations. At the molecular level, however, the dissimilar biological response of the worms in terms of different endpoints such as activity of antioxidant enzymes, lipid peroxidation etc., to varying concentrations of TiO_2 nanoparticles were attributed to agglomeration of nanoparticles which could have influenced particle size and uptake characteristics.

Methodology employed for toxicity assessment:

Filter paper test and artificial soil test (OECD 207 guidelines, 'Earthworms, Acute toxicity test') were adopted in the existing studies. The medium of exposure in the two tests such as filter paper and artificial soil would have varying influences on the physicochemical characteristics of TiO_2 nanoparticles and hence toxicity of the particles. The medium characteristics would also influence bioavailability of the nanoparticles to the worms. However, in the study by Lapied *et al.,* (2011) the authors suggest that there is no significant difference between bioavailability of TiO_2 nanoparticles in soil and water. Another important factor to be taken into consideration would be the method of application of TiO_2 nanoparticles to the exposure medium. The methods of application observed in these studies were (i) dissolution of the particles in solvent and introduction into exposure media (ii) introduction of the nanopowder/nanocomposite directly as dry powder into the exposure media. This could also have an influence on the bioavailability of the nanoparticles to the worms. Hence all these parameters should be taken into account for ecotoxicity assessment of TiO_2 nanoparticles on earthworms in order to develop a comprehensive and holistic risk assessment framework for TiO_2 nanoparticles in the terrestrial ecosystem. The current study aims to address the relation between particle size, an important physicochemical parameter of TiO_2 nanoparticles and the toxicity exerted on earthworms. Effect of TiO_2 (rutile) nanoparticles on the earthworm *Eisenia fetida* was studied using filter paper contact method (acute toxicity test method, OECD 207 guidelines) for a duration of 2 days and endpoints such as activity of antioxidant enzymes (SOD, CAT & GR) and lipid peroxidation were assessed. The antioxidant enzymes chosen have basal level activities in the cell as they maintain the redox balance in the cells. However when there is additional stress to the cell, the activity levels of these enzymes fluctuate depending upon the degree of stress encountered. According to previous studies on effect of TiO_2 nanoparticles on earthworms (Hu *et al.,*2010; Canas *et al.,*2011; Lapied *et al.,* 2011 and McShane *et al.,*2012), the main mechanism of toxicity of these nanoparticles is the induction of oxidative stress by the generation of reactive oxygen species (ROS). Under such conditions, there is variation in the activity levels of the antioxidant enzymes when compared with control. The activity levels increase or decrease depending upon the enzyme type and concentration of the nanoparticles. From previous studies it is evident that there is no clear dose-response relationship with regard to enzyme activity and concentration observed in the test organism. The previous studies also attribute to varying pattern of enzyme activity of the antioxidant system consequently becoming overwhelmed at increasing concentrations of the nanoparticles. This is also supported by the difference in malondialdehyde (MDA) content which is an indicator of lipid peroxidation that usually occurs when the antioxidant defenses are overwhelmed and the redox balance in the cell is disrupted by ROS. Results of the current study are consistent with the above findings. An important parameter that has not been addressed in the previous studies is the effect of particle size rather than concentration of the nanoparticles on various endpoints. This has been addressed in the current study. The current study also reiterates the induction of oxidative stress in the test organism by TiO_2 (rutile) nanoparticles in comparison with the control, by studying activity profiles of three antioxidant enzymes (CAT, GR, SOD) and the extent of lipid peroxidation.

2.9 Highlighting the importance of bio-monitoring and risk assessment of ENPs

The global sales of nano-based products are expected to reach 450 billion euros within the next year (Hanssen *et al.*, 2008). First generation nano products are already in the market and virtually dominate almost every sector of the commercial setting. Consequently, release of these particles into the environment and subsequent exposure to human populations and natural biota is inevitable and already observable in some cases (Fabrega *et al.*, 2011; Nowack *et al.*, 2011). However, the risk assessment strategies for these emerging pollutants are not well defined and are based on traditional chemical risk assessment strategies without taking into consideration their unique properties compared to their bulk counterparts. It is also these unique properties that raise concern regarding the unforeseen environmental and toxicological consequences (Colvin., 2003; RCEP., 2008). The main challenge with these pollutants is the lack of adequate knowledge on how to detect and quantify them in various environmental matrices. This depends on understanding the inherent behavior of these particles in various exposure scenarios, which in turn is dictated by the physicochemical characteristics of the ENPs. The essential characterization parameters have been identified by a number of authors (Oberdorster *et al.,*2005; Klaine *et al.,* 2008; Powers *et al.,* 2006, 2007; Warheit *et al.,* 2008 and Domingos *et al.,* 2009) but they differ in what parameters are considered essential and what are not (Stone *et al.,*2010). A recent review by Michala. E. Pettitt and Jamie Lead (2013) underscores the importance of three issues that need to be addressed for accurate characterization of nanoparticles. They are defining nanomaterials in terms of size, core properties, surface properties, aggregation and solubility characteristics, accounting for variability in nanomaterial preparation and dynamics of nanomaterial in various environmental matrices. This would aid in accurate and reliable characterization of the particles. The next step in risk assessment framework would be to assess the impact of these pollutants on organisms representative of various ecosystems. Interaction of these pollutants with the organisms would vary temporally and spatially and hence would aid in accurate quantification of the effects of ENMs in various exposure scenarios. However, this is a farfetched benchmark for risk assessment as it is interdependent on particle characteristics, fate, transport, behavior, bioavailability in the environment. Other crucial factors to consider would be the effect of pH, ionic strength, nature of electrolytes, presence of organic acids etc. on the particles as they could alter particle characteristics. In order to understand the interdependent effects of nanoparticles and their interacting matrices, sensitive endpoints such as variation in the activity of antioxidant enzymes, lipid peroxidation, genotoxicity, apoptosis, mitochondrial damage etc., in the model organisms should be considered as potential biomarkers. Such studies already exist for various ecosystems but there are gaps that have to be addressed by future research.

In this regard, ecotoxicity data which considered sensitive endpoints for TiO_2 nanoparticles, the most commercially exploited metal oxide nanoparticle (the focus of this review) are available for the aquatic ecosystem. Similar type of data is very scarce for the terrestrial ecosystem which is one of the major sinks for these particles. There are few documented studies involving three types of invertebrates (oligochaetes, isopods and nematodes) representative of the terrestrial ecosystem that are available. This situation highlights the urgency and need for exposure assessment data on ecotoxicity of TiO_2 nanoparticles in the terrestrial ecosystem.

Due to the extensive uncertainties and knowledge gaps that hinder holistic and reliable risk assessment process, several alternative tools for approximate risk assessment have been developed (Khara Greiger., 2010). Some examples of the tools are,

- Precautionary Matrix (Hock *et al.,* 2008): To identify potentially dangerous applications within production

- Categorization framework (Hansen *et al.,* 2008): To identify different exposure potentials based on the location of the nano-structured material in the product

- Multi-Criterion Decision Analysis (MCDA) (Linkov *et al.,* 2007): To compare and rank alternatives in nanomaterial risk decision making

In conclusion, there is an urgent need for the establishment of risk assessment framework for ENMs that encompasses all the requirements and that which is adoptable in time and is accurate and reliable.

2.10 Objectives of the study

- To evaluate the toxicity of TiO_2 rutile nanoparticles in earthworm *Eisenia fetida* by modified Filter paper contact test

- To study activity profiles of three antioxidant enzymes such as catalase (CAT), glutathione reductase (GR) and superoxide dismutase (SOD) when exposed to different concentrations of TiO_2 rutile nanoparticles

- To correlate oxidative stress generated in the worm to particle size of TiO_2 rutile nanoparticles

Chapter 3: Materials and Methods

3.1 Chemicals

Chemicals used in this study were of analytical grade purchased from Sigma Aldrich.

3.2 Test substance

Titanium dioxide rutile nanopowder was purchased from Sigma Aldrich (Titanium (IV) oxide, rutile, <100nm, 99.5% trace metals).

3.3 Test organism

Earthworms (*Eisenia fetida*) were purchased from 'Vermiculture Project', Hyath Nagar, Hyderabad, Andhra Pradesh. They were carefully brought to the laboratory within 2h along with the moist soil. Before testing, these worms were acclimated to test conditions in open space in the vicinity of laboratory in heap (36×18×24 inches.) containing 4 different layers of uncontaminated red soil at the bottom (base soil) 4 inches, a thin layer of leaves, 16 inches of mixed cow dung with soil in a 1:1 ratio, and a thin layer of dried grass on top. Wet gunny bags were used to cover the heaps.

3.4 Physicochemical characterization of TiO_2 rutile nanoparticles

3.4.1 Preparation of nanoparticle suspension

40ml of five different concentrations viz., 0.05, 0.1, 0.15, 0.2 and 0.25mg/ml of TiO_2 rutile nanoparticle suspensions were prepared using ultra-filtered water (pH 7.4, filtered with 0.22μ Whatman filter paper) in 50mL glass beakers (ASGI). Each suspension was subjected to probe sonication (22mM probe, Sonics Vibra Cell) for 30 minutes at 90% amplitude with 30 second 'ON' pulse and 10 second 'OFF' pulse at 750Watts to prevent agglomeration and enhance homogeneity of the dispersion.

3.4.2 Characterization of TiO_2 rutile nanoparticles

3.4.2.1 Characterization by Dynamic Light Scattering (DLS)

TiO_2 rutile nanoparticles were characterized based on their particle size using Malvern Instrument Zetasizer Nano series (Malvern Instruments, Westborough, MA, USA) equipped with a He-Ne laser ($\lambda = 633nm$, max 5mW) and operated at a scattering angle of 173°. It works on the principle of Dynamic Light Scattering also termed as Photon Correlation Spectroscopy or Quasi-Elastic Light Scattering. DLS measures particle size by correlating it with the Brownian motion of the particles in solvent. The velocity of Brownian motion is measured in terms of translational diffusion coefficient from which the hydrodynamic diameter of the particle is calculated by using Stoke's-Einstein equation (DLS technical note, Malvern Instruments).

$d(H) = kT/3\pi\eta D$

where,

$d(H)$ = hydrodynamic diameter

D = translational diffusion coefficient

k = Boltzmann's constant

T = absolute temperature

η = viscosity

The particle size distribution is represented as a temporally dynamic plot between Z average particle size (nm) and percent intensity. Width of the particle size distribution is represented in terms of polydispersity index (PDI).

About 1mL of each concentration of TiO_2 rutile nanoparticle suspension was added into 10mm X 10mm quartz cuvette and the sample was loaded into the Zetasizer for particle size measurement by DLS.

3.4.2.2 Characterization by Scanning Electron Microscopy (SEM)

The particle size in some concentrations was also characterized by Scanning Electron Microscopy (SEM) in order to determine aggregation of the nanoparticles in solvent. SEM works by scanning the sample with a focused beam of electrons that interact with atoms in the sample that are at or near the surface, producing various signals such as secondary electrons (SE), back scattered electrons (BSE), characteristic X-

31

rays, and light (cathodoluminescence) that can be detected and contain information about the sample's topography and composition. The electron beam used is very narrow and hence SEM micrographs have a large depth of field giving rise to high resolution (better than 1nm) images.

SEM characterization was performed to obtain nanoparticle size and morphology on a Hitachi H-7600 tungsten-tip instrument at an accelerating voltage of 100kV. Nanoparticles were examined after suspension in water and subsequent deposition onto formvar/carbon-coated SEM grids. The AMT software for the digital SEM camera was calibrated for size measurements of the nanoparticles. Information on mean size and standard deviation was calculated from measuring over 100 nanoparticles in random fields of view in addition to images that show general morphology of the nanoparticles.

3.5 Exposure of *Eisenia fetida* to TiO_2 rutile nanoparticles

The modified Filter paper test adapted from OECD 207 guidelines for testing of chemicals (Earthworms, Acute Toxicity tests) was chosen for exposure study of TiO_2 rutile nanoparticles in *Eisenia fetida*. The study was conducted on glass petri plates ($68cm^2$ surface area, Borosil) with 10 replicates for each concentration of nanoparticles and 10 for control (distilled water, pH 7.4).The plates were layered with a single layer of Whatman grade I filter sheet. 1.3ml of the suspensions of different concentrations was added to each plate drop-wise till complete wetting of filter paper was observed. Adult worms with well-developed clitellum and having average weights in the range of 200-400mg were washed, dried, weighed (Sartorius) and then introduced into each plate after 6 hours of defecation to eliminate gut contents. The plates were closed and placed on a dark shelf for 48 hours, with intermittent moistening of the filter paper with distilled water (pH 7.4). After 48 hours, the physiological characteristics, behavior, and mortality of worms in each plate were recorded.

S.No	Concentration of TiO$_2$ rutile nanoparticles (mg/cm^2)	No. of worms
1	0.05	10
2	0.1	10
3	0.15	10
4	0.2	10
5	0.25	10
6	Control	10

Table.II. Different concentrations of TiO$_2$ rutile nanoparticles

3.6 Whole body homogenate preparation

The worms in each concentration and control that had survived the exposure were suspended in ice cold 0.1M sodium phosphate buffer (pH 7.4) in the order of 10% of worm weight. The worms in buffer were initially sonicated at 90% amplitude followed by homogenization using mortar and pestle to obtain whole body extract in solution. About 6-10ml of the extract was obtained for each sample in 2ml centrifuge tubes (Tarsons), which were then centrifuged at 10,000g for 10 minutes. The supernatant collected was stored at -20°C.

3.7 Total protein estimation

Total protein content of the whole body homogenate was estimated by Lowry's method (Lowry et al., 1951). Principle of the Lowry's method lies in the reactivity of the peptide nitrogen with the copper (II) ions under alkaline conditions and subsequent reduction of the Folin-Ciocalteau reagent (phosphomolybdic phosphotungstic acid) to heteropolymolybednum blue by the copper catalyzed oxidation of aromatic amino acids especially tyrosine and tryptophan (http://www.bio.davidson.edu). Intensity of the blue color depends upon the tyrosine and tryptophan content of the protein and to a lesser extent cysteine and other residues in the protein. The Lowry's method is sensitive and can detect protein concentrations in the range 0.005-2mg/mL.

Initially, a standard graph was prepared with Bovine Serum Albumin (BSA) (1mg/ml) as protein standard followed by estimation of unknown samples. To 20μL of sample, 200μL of freshly prepared alkaline $CuSO_4$ was added and incubated at room temperature for 10minutes. To this mixture, 20μL of freshly prepared Folin's reagent was added and incubated at room temperature for 30 minutes till a blue colored complex was observed. The samples along with blank were loaded in triplicates in a 96 well plate (NEST) and the absorbance was measured at 600nm using Spectramax M4 (Molecular Devices) spectrophotometer.

3.8 Determination of enzyme activity

Activities of the chosen antioxidant enzymes, catalase, glutathione reductase and superoxide dismutase were estimated in the samples by specific assays described by Claiborne et al., (1985), Schaedle and Bassham (1977), Marklund and Marklund (1974) respectively. The absorbance values of samples in each assay were measured at different wavelengths using Spectramax M4 (Molecular Devices) configured for unit cm path length. The specific activity of the enzymes was finally reported in terms of units/mg of protein/min.

3.8.1 Catalase (EC 1.11.1.6) assay

Principle: The principle of this assay is based on the conversion of hydrogen peroxide into water and molecular oxygen by the enzyme catalase. The activity of catalase is determined indirectly by measuring the change in absorbance of H_2O_2 at 240nm.

Method: To 5μL of the sample, 250μL of 50mM sodium phosphate buffer containing 5mM freshly prepared hydrogen peroxide (H_2O_2) was added with caution to avoid bubble formation. Triplicates for each sample were taken, loaded into a 96 well plate and their absorbance was measured at 240nm in Spectramax M4 for a time period of 2 minutes with 10 second interval.

3.8.2 Glutathione reductase (EC. 1.6.4.2) assay

Principle: The assay is based on the principle of reduction of glutathione disulfide (GSSG) to glutathione (GSH) by the enzyme glutathione reductase with (Nicotinamide Adenine Dinucleotide Phosphate reduced form) NADPH as the electron donor. The activity of the enzyme glutathione reductase is determined indirectly by measuring the change in absorbance of NADPH at 340nm.

Method: 5µL of 20mM oxidized glutathione (GSSG), 10µL of 10mM Ethylene diamine tetra acetic acid (EDTA), 133µL of 0.1M sodium phosphate buffer were added to 50µL of sample. Each sample was taken in triplicates in a 96 well plate and absorbance was measured at 340nm for a time period of 2 minutes with 10 second interval.

3.8.3 Superoxide dismutase (EC. 1.15.1.1) assay

Principle: The assay is based on the principle of auto-oxidation of pyrogallol (the absorption maximum of which is at 420nm) by the superoxide anion radical which is a type of ROS. The enzyme superoxide dismutase rapidly dismutases univalently reduced oxygen into molecular oxygen and hydrogen peroxide. Hence in the reaction mixture, the activity of SOD is measured indirectly by measuring the absorbance of auto-oxidized pyrogallol.

Method: 9µL of 24mM pyrogallol prepared in 10mM HCl, 289µL of 0.1M sodium phosphate buffer were added to 6µL of the sample. The samples were loaded in triplicates in a 96 well plate and the absorbance was measured at 420nm for a time period of 10minutes with 1minute interval.

3.9 Lipid Peroxidation study

Principle: The study was carried out based on the modified protocol of Utley *et al* (1967) for estimating the malondialdehyde (MDA) content. Oxidative degradation of the lipid membrane (lipid peroxidation) caused by reactive oxygen species (ROS) leads to the production of hydroperoxides and aldehydes such as MDA which are collectively termed as TBARS (2-Thiobarbituric acid reactive substances). In the assay, a chromogenic reagent, 2-Thiobarbituric acid reacts with TBARS and yields a chromophore with absorbance maximum at 535nm.

Method: A standard graph was prepared using malondialdehyde (MDA) and then the malondialdehyde content in unknown samples was estimated. In this method, the samples were mixed with 0.5ml incubation mixture (0.04M Tris-HCl, 0.25M KCl, 400µM Ascorbate, 20µM $FeSO_4$). To this mixture, 0.5ml of chilled Trichloroacetic acid and 0.5ml of 0.67% Thiobarbituric acid were added. This mixture was centrifuged at 1000g for 15 minutes in a cooling centrifuge. The supernatant was then collected in fresh tubes and placed in a boiling water bath for 10 minutes till the appearance of pink color. The absorbance of the samples was measured at 535nm by Beckman Coulter spectrophotometer.

Chapter 4: Results and Discussion

4.1 Characterization of TiO$_2$ rutile nanoparticles

4.1.1 Characterization by Dynamic Light Scattering (DLS) and Scanning Electron Microscopy (SEM)

DLS data indicates the size distribution of different concentrations of TiO$_2$ rutile nanoparticles having nominal size (< 100nm) in ultra-filtered water (pH 7.4) as illustrated in Figure 4.

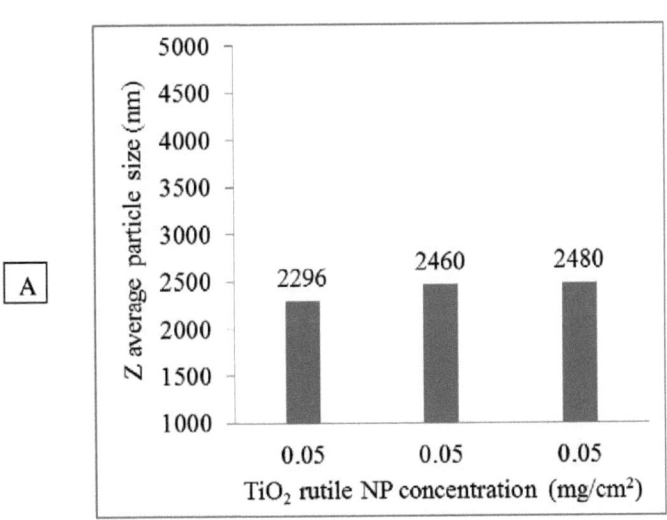

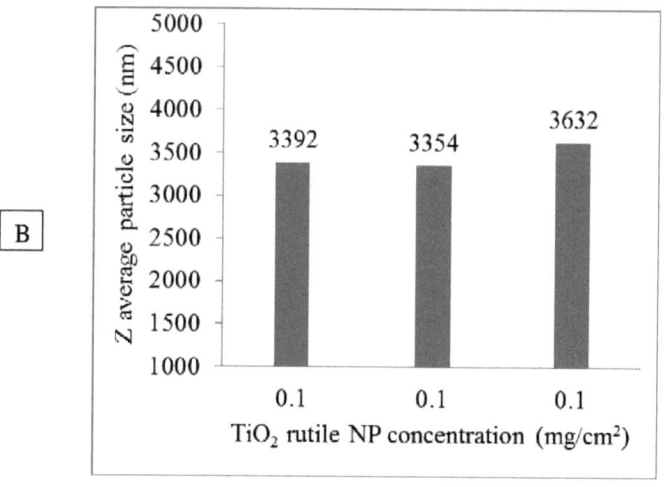

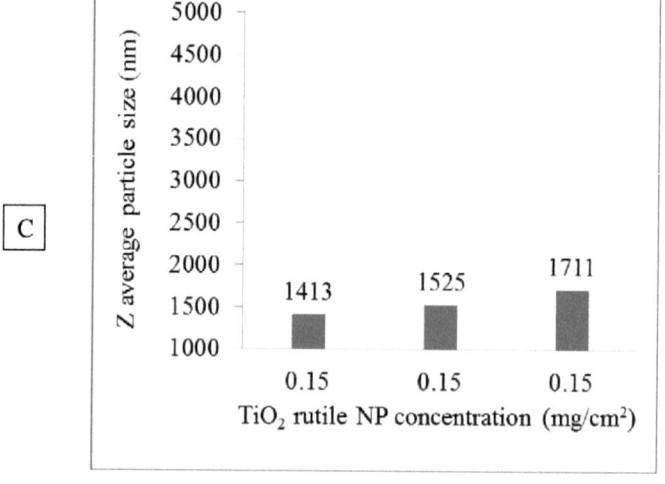

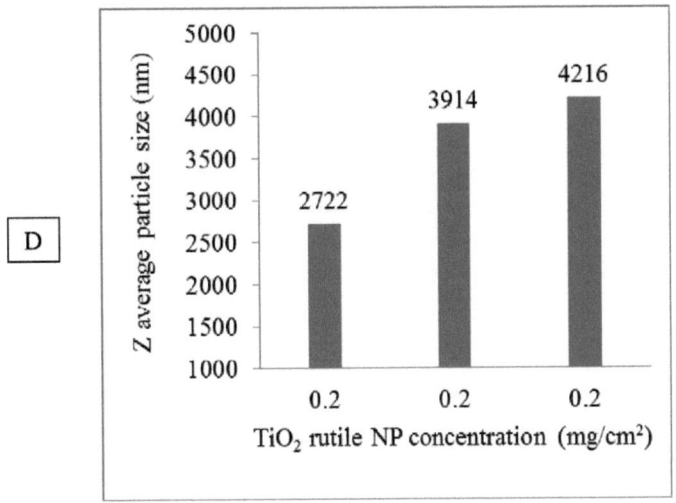

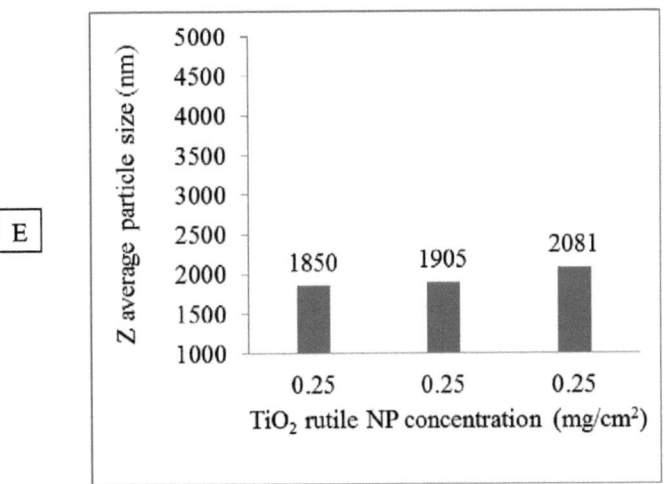

Figure. 4 Particle size distribution in concentrations 0.05mg/mL (A) 0.1mg/mL (B) 0.15mg/mL (C) 0.2mg/mL (D) 0.25mg/mL (E)

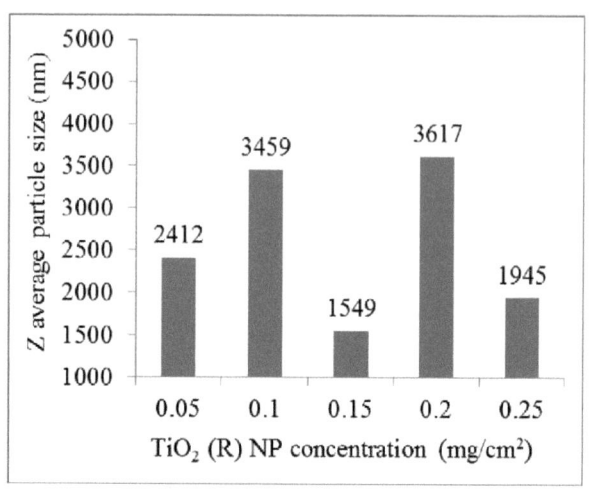

Figure.5 Average particle size distribution at each concentration

From the average particle size distribution, it is evident that there is no linear correlation between increase in concentration and particle size of TiO_2 rutile nanoparticles. The smallest and largest particle sizes correspond to $0.15mg/cm^2$ and $0.2mg/cm^2$ respectively. The variation in particle size (hydrodynamic size) is primarily due to agglomeration and aggregation of nanoparticles in the solvent. When nanoparticles are dispersed in liquids their hydrodynamic size is often greater than the primary particle size or nominal size (Buford *et al.*, 2007; Sager *et al.*, 2007). This is attributed to the agglomeration and aggregation effects of the nanoparticles in aqueous media and is often influenced by factors such as primary particle size, medium components, pH, ionic strength, surface charge of nanoparticles and nature of electrolytes in the medium (Lead., 2009; Keller *et al.*, 2010; Hotze *et al.*, 2010; Stone *et al.*, 2010; Zhou *et al.*, 2010). Agglomeration of TiO_2 nanoparticles has been observed in cell culture media (William Vevers and Awadhesh Jha., 2008; Zhouhir Alloni., 2009; Zhaoxia Ji., 2010; Anna Lankoff *et al.*, 2012) and natural aqueous matrices (Keller *et al.*, 2010; Marina Belen Romanello and Maria Fidalgo de Cortalezzi., 2013). In the current study, agglomeration of TiO_2

rutile nanoparticles in solvent in comparison to their nanopowder form has been confirmed by SEM analysis illustrated in Figure 5. Although primary particle size of the TiO_2 rutile nanoparticles is <100nm, particle size range was found to be in micrometers in water as dispersant (Figure 6A and 6B).

A

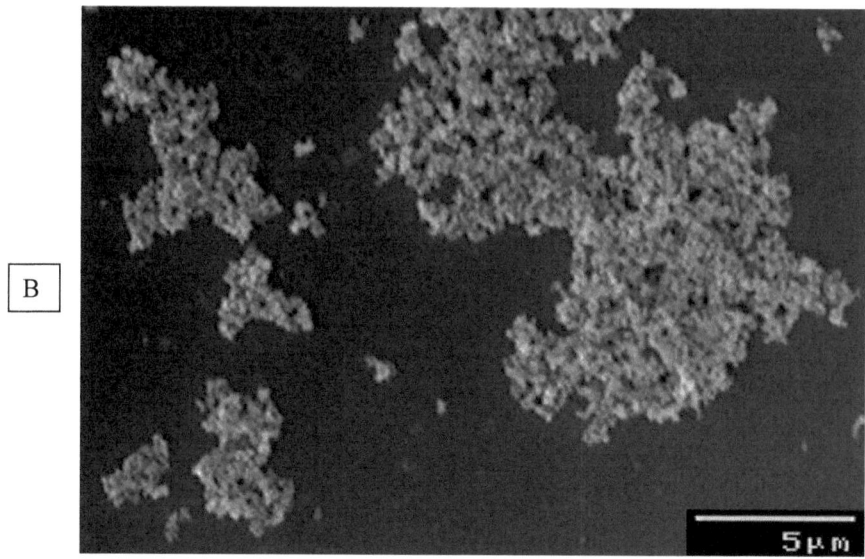

Figure.6 Agglomeration of TiO_2 nanoparticles in powder (A) and in distilled
water (B) captured by Scanning Electron Microscope (JEOL JSM 5400 at 15 kV)

Varying degree of agglomeration of TiO_2 rutile nanoparticles can also be influenced
by particle size, an important physicochemical parameter of nanoparticles that needs
to be addressed in nanotoxicological studies (Oberdorster *et al.*, 2005; Powers *et al.*,
2006). In the current study, a mixture (<100nm) of TiO_2 rutile nanoparticles have
been considered for their ecotoxicity assessment on earthworms and hence could
correspond to varying degree of agglomeration at different concentrations. The
agglomeration process of TiO_2 nanoparticles for diverse morphologies appears to be
controlled by electrostatic and van der Waal's interactions (Guzman., 2006; Thio.,
2010; Hanaor., 2011; Liu., 2011) which is correlated to particle size and surface
charge density. According to the study done by Abbas *et al.*, (2008), increase in
surface charge occurs as nanoparticle size decreases and the increase in particle
surface charge would augment electrostatic repulsive forces, mask agglomeration,
improve dispersity and reduce hydrodynamic diameter (Suttiponparnit *et al.*, 2011).

The current study, however, employed a mixture of TiO_2 rutile nanoparticles having varying sizes each of which would correspond to different surface charge densities and hence exhibit varying levels of agglomeration at different concentrations in solvent. Difference in the degree of agglomeration influences the average hydrodynamic particle size which is evident from the DLS data.

4.2 Exposure of *Eisenia fetida* to TiO_2 rutile nanoparticles

After 48 hours of exposure on filter paper containing different concentrations (0.05, 0.1, 0.15, 0.2, 0.25mg/cm²) of TiO_2 rutile nanoparticle suspension, the percent morality of worms in each concentration and control are given below.

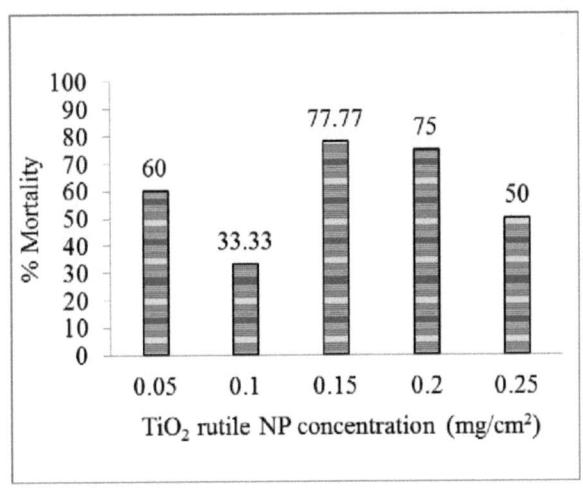

Figure.7 Mortality of worms exposed to TiO_2 rutile nanoparticles

The highest (77.77%) and lowest (33.33%) percent mortality rates correspond to 0.15mg/cm² and 0.1mg/cm² respectively. The particle sizes corresponding to these mortality rates are 1549nm and 3459nm, which are the least and second highest particle sizes respectively. Earthworms exposed to 1549nm nanoparticles have exhibited the highest percent mortality and those exposed to 3459nm have exhibited

43

the lowest mortality percentage. However the mortality rate does not seem to correspond to particle size nor concentration after $0.15mg/cm^2$. This indicates that beyond the concentration range of $0.15mg/cm^2$, there is uncertainty in the discernible correlation between particle size of TiO_2 rutile nanoparticles and mortality of earthworms. However the influence of particle size on mortality of the worms rather than the effect of concentration of TiO_2 rutile nanoparticles on mortality of the worms is evident in the concentration range of $0.05mg/cm^2$-$0.15mg/cm^2$. In the study done by Canas et al (2011), toxicity of TiO_2 anatase nanoparticles (32nm) to the earthworm $Eisenia$ $fetida$ was studied by filter paper contact method. The authors had observed agglomeration of nanoparticles in DLS and SEM measurements, in a particular concentration range (1-10 mg/L). They had also reported that there was no clear relation between concentration of TiO_2 nanoparticles and mortality of the worms (i.e., no clear dose - response was seen). This study is in coherence with the current study in that a clear correlation between concentration of nanoparticles and mortality of the worms could not be established. This condition could be attributed to the influence of some of the physicochemical properties such as particle size and agglomeration effects. Also, in the study by Canas et al, the influence of particle size on mortality was not established. The current study is contradictory to the study carried out by Lovern et al (2007) on the behavioural and physiological changes in $Daphnia$ $magna$ when exposed to TiO_2 nanoparticle suspensions. The authors had observed an increase in mortality with increase in concentration of TiO_2 nanoparticles. The suspensions with TiO_2 nanoparticles used for the toxicity study were stabilized to maintain the particle size in the range of 10-20nm and also filtered. This would have eliminated the formation of nanoparticle agglomerates with large hydrodynamic diameters. The resulting uniformly sized nanoparticles in suspension were finally used for the ecotoxicity assessment on $D.magna$. The difference between this study and the current study lies in the application of a stabilizer Tetrahydrofuran (THF) for maintaining uniform particle size and filtration of the nanoparticle suspension to remove agglomerates. This could have aided in the establishment of a linear correlation between increase in concentration of TiO_2 nanoparticles and mortality of *D.magna*. However the relation between particle size and mortality of *D.magna* was not addressed in this study.

The functionality and biological effects of Titania are controlled by its physicochemical properties (Suttiponparnit *et al.,* 2011). Physicochemical properties such as particle size, agglomeration state, surface charge etc., tend to be altered once the nanoparticles are introduced into any biological (test) medium and these variations may influence the toxicological response of the organism exposed to the nanoparticles (Powers *et al.,* 2007). The nanoparticle size governs its interactions with biological systems including absorption, distribution, metabolism and excretion (Renwick *et al.,* 2001; Borm *et al.,* 2006; Choi *et al.,* 2007). The phenomenon of agglomeration of nanoparticles is interdependent on particle size and surface charge density and also tends to influence nanoparticle reactivity (Waychunas *et al.,* 2005; Gilbert., 2009; Zeng., 2009) and response of the test organism upon exposure (Hoshino *et al.,* 2004; Lockman., 2004; Choi *et al.,* 2007; Wu., 2010). Uptake of nanoparticles by the worms depends on particle size. Many studies have reported that small sized TiO_2 nanoparticles have serious consequences than large particles in exposed organisms. In the study by Hunde-Rinke *et al* (2006), immobilization of invertebrates was more serious when small-sized TiO_2 nanoparticles were used. However, from the current study, it is evident that smaller particle size of TiO_2 rutile nanoparticles were taken up much easily by the worms in comparison to the larger particle agglomerates. This can be directly correlated to the degree of stress encountered by the worms that were exposed to the least and highest particle sizes. In the current study, activities of 3 antioxidant enzymes catalase (CAT), glutathione reductase (GR) and superoxide dismutase (SOD) and lipid peroxidation in the worms exposed to TiO_2 rutile nanoparticles were examined in order to determine the

45

induction and extent of oxidative stress in the worms and their correlation to nanoparticle size.

4.3 Total protein estimation

The protein concentration in the whole body homogenate of control and treated worms was determined from their corresponding absorbance values at 600nm by solving the equation $y = 0.079 + 1.025(x)$, where x and y are concentration (mg/ml) and absorbance values respectively.

4.4 Determination of enzyme activity and lipid peroxidation

A cell is normally subjected to oxidative stress during various metabolic reactions in which oxygen serves as the major oxidant. Oxidative stress results due to the imbalance between pro-oxidative and anti-oxidant states that result due to the production of reactive oxygen species (ROS) such as superoxide anion, hydrogen peroxide, hydroxyl radicals etc. in the cell. The equilibrium between these two states is restored in case of normal aerobic metabolism. However, when the cell faces additional oxidative stress, the balance tilts towards the pro-oxidative state resulting in the damage to major biomolecules such as proteins, lipids, carbohydrates and nucleic acids. The balance is usually maintained by enzymatic and non-enzymatic factors. The enzymatic factors include antioxidant enzymes such as catalase, glutathione reductase, glutathione peroxidase, NADPH oxidase, superoxide dismutase etc. The non-enzymatic factors would include vitamins, molecules like glutathione etc. Activities of antioxidant enzymes, in particular, tend to fluctuate when a higher degree of oxidative stress is encountered. Also the loss of membrane integrity due to lipid peroxidation caused by ROS, is also a strong indicator of oxidative stress. The lipid peroxidation levels in this study were estimated based on the standard graph prepared for malondialdehyde (MDA). The MDA content in the whole body homogenate of worms exposed to TiO_2 rutile nanoparticles and control were calculated from the equation, $y = 0.0051(x) + 0.00075$. Therefore activities of three antioxidant enzymes and levels of lipid peroxidation have been chosen in this study to investigate the induction of oxidative stress by TiO_2 rutile nanoparticles in earthworm *Eisenia fetida*.

From the study it is evident that there is induction of oxidative stress in the earthworms exposed to TiO_2 rutile nanoparticles. It is also evident from the differences in antioxidant enzyme activities and lipid peroxidation in the exposed worms and control. Many studies documented the induction of oxidative stress by TiO_2 nanoparticles in organisms such as *Cyprinus carpio* (Hao *et al.*, 2009), *Mytilus galloprovincialis* (Canesi *et al.*, 2010), *Haliotis diversicolor supertexta* (Xiaoshan *et al.*, 2011), *Eisenia fetida* (Hu *et al.*, 2010; Canas *et al.*, 2011; McShane *et al.*, 2012), *Caenorhabditis elegans* (Quili *et al.*, 2012), and in a number of in vitro models (Limbach *et al.*, 2007; Jana Petković *et al.*, 2011; Pujalte *et al.*, 2011; Shukla *et al.*, 2011). Although the toxicity of TiO_2 nanoparticles was assessed in various organisms, the relation between particle size and oxidative stress was not explained. However in the study by Jana Petkovic *et al* (2011) it has been highlighted that smaller nanoparticles can induce greater oxidative stress. This is quite evident from the current study in terms of mortality and activity of enzymes corresponding to particle size 1549 nm at 0.15mg/cm^2 concentration. Also the degree of oxidative stress varies across concentrations but a clear dose - response relation cannot be established. This could be attributed to the particle size variations across concentrations that seem to be influenced by agglomeration effects. Accordingly varying levels of activity of antioxidant enzymes and lipid peroxidation are evident across concentrations (0.05, 0.1, 0.15, 0.2, 0.25mg/cm²). This is illustrated in Figure.8

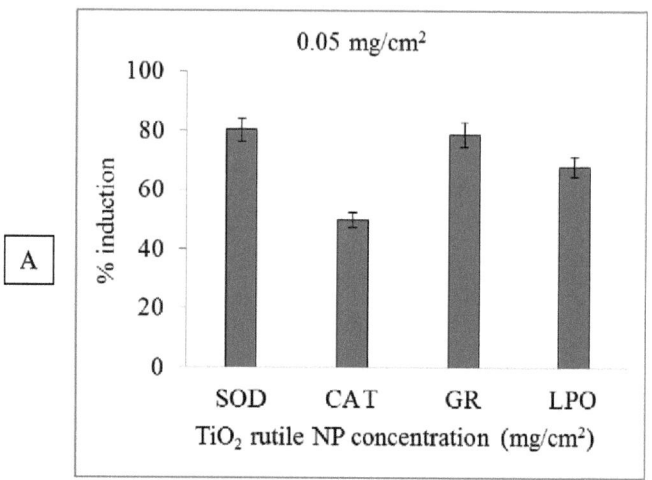

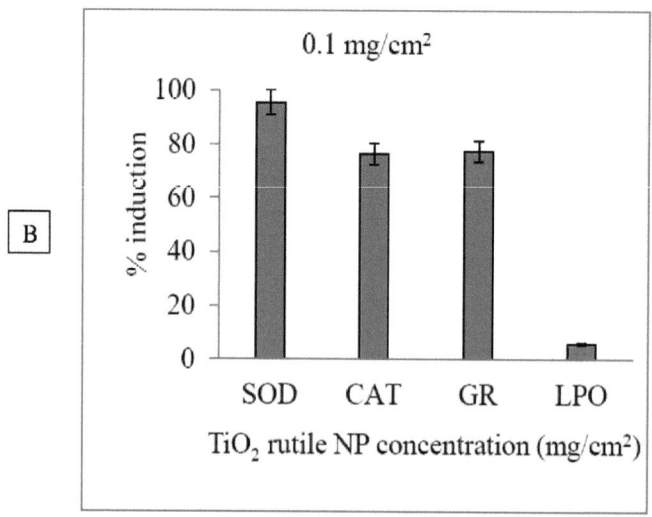

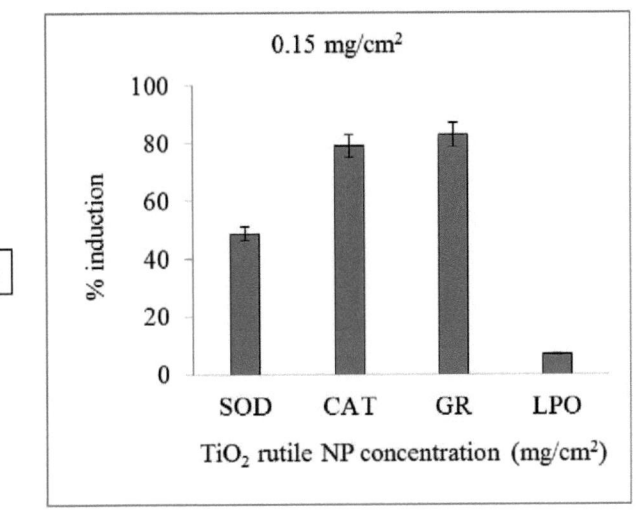

C

0.15 mg/cm²

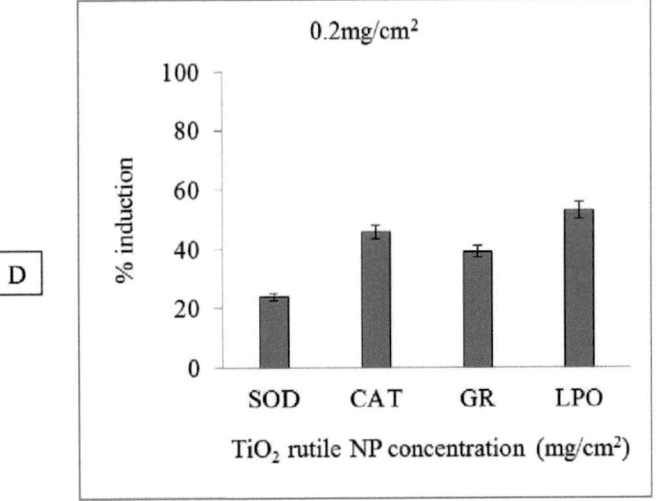

D

0.2mg/cm²

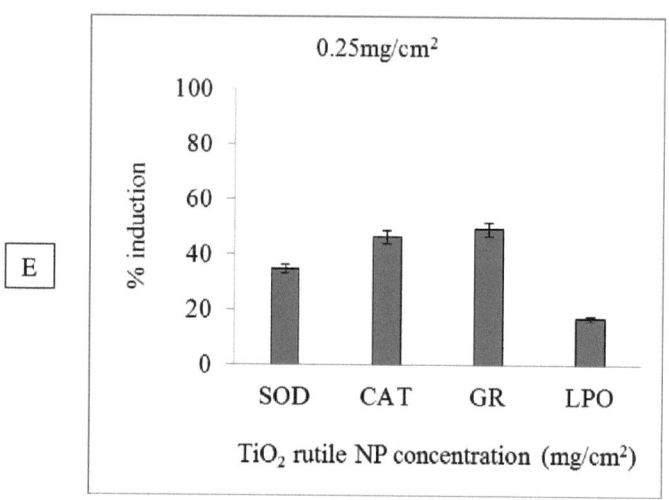

Figure.8 Variation in enzyme activities and lipid peroxidation at different
concentrations of TiO$_2$ rutile nanoparticles

Comprehensive outlook on the activity of three different antioxidant enzymes and
corresponding degree of lipid peroxidation has been represented in Figure 9.

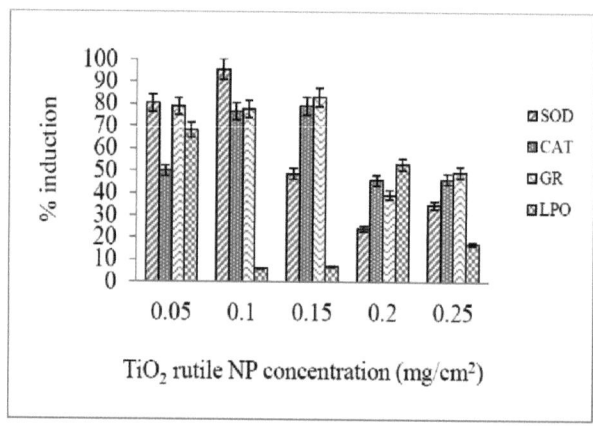

Figure.9 Activity profiles of 3 different antioxidant enzymes and lipid
peroxidation levels at 5 different concentrations

At $0.1mg/cm^2$ and $0.15mg/cm^2$, activity of all three enzymes has been induced significantly. Consequently, induction of lipid peroxidation is less at these concentrations. This indicates that the oxidative stress induced by the nanoparticles was counteracted by the cell and the consequent damages to biomolecules such as lipids were mitigated. In the concentrations $0.05mg/cm^2$ and $0.2mg/cm^2$, the level of lipid peroxidation seems to be high despite increase in activity of antioxidant enzymes. This could be because at these concentrations, the antioxidant defense system could have been overwhelmed by oxidative stress. A similar observation has been reported by Hu *et al* (2010) in his study on the toxicity of TiO_2 nanoparticles to *Eisenia fetida*. Also he has reported the correlation between decrease in antioxidant enzyme activity (SOD and CAT) and increase in levels of lipid peroxidation. In the study by Valant *et al* (2012) on the effect of TiO_2 nanoparticles on the digestive gland membrane of terrestrial isopod, *Porcellio scaber,* lipid peroxidation was detected after long exposure period. In another study by Hao *et al* (2009), the decrease in the levels of antioxidant enzymes such as SOD, CAT and peroxidase (POD) correlated with increased levels of lipid peroxidation. This is in coherence with the current study, where there is induction of lipid peroxidation by TiO_2 rutile nanoparticles and also the increased levels of lipid peroxidation correspond to decreased activity of antioxidant enzymes.

In the current study, fluctuations are observed in terms of enzyme activity and lipid peroxidation levels at concentrations $0.1mg/cm^2$ and $0.15mg/cm^2$. These concentrations correspond to the second highest and least particle sizes and lowest and highest mortality rates. Hence a difference in the response of the antioxidant enzymes and consequent degree of lipid peroxidation is expected. But from the results, it is quite evident that there is no significant difference in these parameters at these two concentrations. This could be attributed to the variations in particle size distributions at these concentrations which in turn is dependent on individual particle size and agglomeration effects. Since a mixture of particles with varying sizes having different surface charge densities were used, the degree of agglomeration would have varied significantly. This would have influenced the uptake of particles by the worms. In addition, the varying surface charges of differently sized particles and agglomerates would have influenced their reactivity with the biological system of the worms.

Hence from the current study, it is evident that the varied size of TiO_2 nanoparticles exhibit agglomeration in water dispersion thereby influencing its toxic effects on *Eisenia fetida*. TiO_2 nanoparticles can also cause oxidative stress in earthworms. However further studies are warranted to study the correlation between various physicochemical parameters (particle size, surface properties, agglomeration characteristics and mechanisms, solubility characteristics, variability in ENM preparation, behavior and interaction of ENMs in various matrices) and their influence on toxicity in living systems.

Conclusion

From the study on the toxicity of rutile titanium dioxide nanoparticles on earthworm *Eisenia fetida*, the following insights were gained.

- A discernible correlation could be observed between particle size and mortality

- The particle size influenced the agglomeration characteristics of the nanoparticles which in turn influenced mortality of the worms

- The correlation between particle size and mortality was not discernible at nanoparticle concentrations beyond $0.15mg/cm^2$

- TiO_2 rutile nanoparticles are indeed capable of inducing oxidative stress in the worms as it is evident from mortality rates, antioxidant enzyme activities and lipid peroxidation.

- Activities of the enzymes seem to fluctuate with variations in particle size of the nanoparticles

Recommendations

❖ Characterization study

- Complete physicochemical characterization of TiO_2 nanoparticle in terms of particle size distribution, agglomeration characteristics, zeta potential, and the effect of sonication type, time and amplitude on particle characteristics are to be carried out before exposure studies
- The effect of TiO_2 nanoparticle type (anatase/rutile and their mixtures) on toxicity to the organism should also be considered

❖ Exposure study

- Particle diameter in various media should be measured
- The nanoparticle concentration present on the filter paper in filter paper contact method should be quantified before and after exposure
- The quantity of uptake of particles in the test organism should be determined by methods such as inductively coupled plasma-mass spectrophotometry (ICP-MS)

❖ General

- The validation of existing methods so as to determine LC_{50} of TiO_2 nanoparticles in various indicator organisms is required
- In future studies the effect of physicochemical properties on toxicity of TiO_2 nanoparticles is warranted
- The particle - particle interactions and the consequent effects of agglomeration need to be better understood
- DNA damage, another sensitive indicator of toxicity should also be studied by genotoxicity assay

References

1. A.A.Keller, H.Wang, D.Zhou, H.S.Lenihan, and G.Cherr, 'Stability and aggregation of metal oxide nanoparticles in natural aqueous matrices, 'Environmental Science and Technology 44: 1962–1967, 2010.

2. A.Balamurugan, G.Sockalingum, J.Michel, J.Faure, V.Banchet, L.Wortham, 'Synthesis and characterization of sol gel derived bioactive glass for biomedical applications', Materials Letters 60 (29–30): 3752–7, 2006.

3. A.Boxall, K.Tiede, and Q.Chaudhry, 'Engineered nanomaterials in soils and water: how do they behave and could they pose a risk to human health', Nanomedicine 2: 919-927, 2007.

4. A.Claiborne, 'Catalase activity', In CRC Handbook of Methods for Oxygen Radical Research, ed. Greenwald RA.CRC Press, Inc., Boca Raton, FL, pp. 283–284. 1985.

5. A.G.Cattaneo, R.Gornati, M.Chiriva-Internati, and G.Bernardini, 'Ecotoxicology of nanomaterials: the role of invertebrate testing', Invertebrate Survival Journal 6: 78-97, 2009.

6. A.Gourmelon and J.Ahtiainen, 'Developing test guidelines on invertebrate development and reproduction for the assessment of chemicals, including potential endocrine active substances—the OECD perspective', Ecotoxicology 16: 161–167, 2007.

7. A.Hoshino, K.Fujioka, T.Oku, M.Suga, Y.F.Sasaki YF, T.Ohta, M.Yasuhara, K.Suzuki, and K.Yamamoto, 'Physicochemical properties and cellular toxicity of nanocrystal quantum dots depend on their surface modification', Nano Letters, 4 (11): 2163-2169, 2004.

8. A.Kahru and H.C.Dubourguier, 'From ecotoxicology to nanoecotoxicology', Toxicology 269: 105-119, 2010.

9. A.Lankoff, W.J.Sandberg, A.Wegierek-Ciuk, H.Lisowska, M.Refsnes, B.Sartowska, P.E.Schwarze, S.Meczynska-Wielgosz M.Wojewodzka, and M.Kruszewski, 'The effect of agglomeration state of silver and titanium dioxide nanoparticles on cellular response of HepG2, A549 and THP-1 cells', Toxicology Letters 208 (3): 197-213, 2012.

10. A.Nel, T.Xia, L.Madler, N.Li, 'Toxic potential of materials at the nano level', Science 311: 622-627, 2006.

11. Alex Weir, Paul Westerhoff, Lars Fabricius, Kiril Hristovski, and Natalie von Goetz, 'Titanium Dioxide Nanoparticles in Food and Personal Care Products', Environmental Science and Technology 46 (4): 2242-2250, 2012.

12. Anja Menard, Damjana Drobne, and Anita Jemec, 'Ecotoxicity of nanosized TiO_2-Review of in vivo data', Environmental Pollution 159: 677-684, 2011.

13. B.Gilbert, R.K.Ono, K.A.Ching, and C.S.Kim, 'Effect of nanoparticle aggregation processes on aggregate structure and metal uptake', Journal of Colloid and Interface Science, 339: 285-295, 2009.

14. B.J.R.Thio, D.X.Zhou, and A.A.Keller, 'Influence of natural organic matter on the aggregation and deposition of titanium dioxide nanoparticles', Journal of Hazardous Materials 189: 556–563, 2011.

15. B.Nowack and T.D.Bucheli, 'Occurrence, behaviour and effects of nanoparticles in the environment', Environmental Pollution 150: 5–22, 2007.

16. B.Nowack, H.F.Krug, and M.Height, '120 Years of Nanosilver History: Implications for Policy Makers', Environmental Science & Technology 45(4): 1177-1183, 2011.

17. B.Wu, R.Huang, M.Sahu, X.Feng, P.Biswas, and Y.J.Tang, Science of the Total Environment, 408: 1955, 2010.

18. C.O.Robichaud, A.E.Uyar, M.R.Darby, L.Zucker, and M.Wiesner, 'Estimates of upper bounds and trends in nano-TiO_2 production as a

basis for exposure assessment', Environmental Science and Technology 43: 4227-4233, 2009.

19. C.W.Hu, M.Li, Y.B.Cui, D.S.Li, J.Chen, and L.Y.Yang, 'Toxicological effects of TiO_2 and ZnO nanoparticles in soil on earthworm *Eisenia fetida*', Soil Biology and Biochemistry 42: 586–591, 2010.

20. D.J.Spurgeon, C.Svendsen, P.K.Hankard, J.M.Weeks, P.Kille, and S.K.Fishwick, 'Review of sublethal ecotoxicological tests for measuring harm in terrestrial ecosystems', R&D Technical Report P5-063/TR1, Environment Agency, Bristol, 2002.

21. D.Reyes-Coronado, G.Rodrıguez-Gattorno, M.E.Espinosa-Pesqueira, C.Cab, R.de Coss, and G.Oskam, 'Phase-pure TiO_2 nanoparticles: anatase, brookite and rutile', Nanotechnology 19: 10-19, 2008.

22. D.Zhou and A.A.Keller, 'Role of morphology in the aggregation kinetics of ZnO nanoparticles', Water Research 44: 2948–2956, 2011.

23. DA.H.Hanaor, and C.C.Sorrell, 'Review of the anatase to rutile phase transformation', Journal of Material Sciences 46: 855–874, 2011.

24. David.B.Warheit, 'How meaningful are the results of Nanotoxicity Studies in the absence of adequate material characterization?', Toxicological Sciences 101 (2): 183-185, 2008.

25. E.Lapied, J.Y.Nahmani, E.Moudilou, P.Chaurand, J.Labille, J.Rose, J.M.Exbrayat, D.H.Oughton and E.J.Joner, 'Ecotoxicological effects of an aged TiO_2 nanocomposite measured as apoptosis in the anecic earthworm *Lumbricus terrestris* after exposure through water, food and soil', Environment International 37: 1105–1110, 2012.

26. E.M.Hotze, T.Phenrat, and G.V.Lowry, 'Nanoparticle aggregation: challenges to understanding transport and reactivity in the environment', Journal of Environmental Quality 39: 1909–1924, 2010.

27. E.Navarro, A.Baun, R. Behra, N.B.Hartmann, J.Filser, A.J. Miao, A.Quigg, P.H.Santschi, and L.Sigg, 'Environmental behavior and

ecotoxicity of engineered nanoparticles to algae, plants, and fungi', Ecotoxicology 17: 372–386, 2008.

28. EPA, U. External Review Draft - Nanomaterial Case Studies: Nanoscale Titanium Dioxide in Water Treatment and in Topical Sunscreen. National Center for Environmental Assessment, Office of Research and Development, U.S. Environmental Protection Agency; Research Triangle Park, NC: p. 222EPA/600/R-09/057, 2009.

29. G.A.Waychunas, C.S.Kim and J.F.Banfield, 'Nanoparticulate iron oxide minerals in soils and sediments: unique properties and contaminant scavenging mechanisms', Journal of Nanoparticle Research 7: 409–433, 2005.

30. Günter Oberdörster, Andrew Maynard, Ken Donaldson, Vincent Castranova, Julie Fitzpatrick, Kevin Ausman, Janet Carter, Barbara Karn, Wolfgang Kreyling, David Lai, Stephen Olin, Nancy Monteiro-Riviere, David Warheit, and Hong Yang, 'Principles for characterizing the potential human health effects from exposure to nanomaterials: elements of a screening strategy', Particle and Fibre Toxicology, 2:8, 2005.

31. H.C.Utley, F.Bernheim, and P.Hachslein, 'Effect of sulfhydryl reagent on peroxidation in microsome', Archives of Biochemistry and Biophysics 260: 521-531, 1967.

32. H.McShane, M.Sarrazin, J.K.Whalen W.H.Hendershot, and G.I.Sunahara, 'Reproductive and behavioral responses of earthworms exposed to nano-sized titanium dioxide in soil', Environmental Toxicology and Chemistry 31:184–193, 2012.

33. H.S.Choi, W.Liu, P.Misra, E.Tanaka, J.P.Zimmer, B.I.Ipe, M.G.Bawendi, and J.V.Frangioni, 'Renal clearance of quantum dots', Nature Biotechnology, 25:1165-1170, 2007.

34. H.Zeng, A.Singh, S.Basak, K.U.Ulrich, M.Sahu, P.Biswas, J.G.Catalano, and D.E.Giammar, 'Nanoscale size effects on Uranium

(VI) adsorption to Hematite', Environmental Science and Technology, 43: 1373-1378, 2009.

35. http://nice.asu.edu/mechanism/passive-nanostructure

36. http://www.bio.davidson.edu

37. http://www.epa.gov

38. http://www.nano.monash.edu.au/case-studies/terry-turney.html

39. http://www.nanotechproject.org/cpi

40. http://www.rathenauinstituut.com/downloadfile.asp?ID=1486

41. http://www.rsc.org/

42. http://www.unep.org

43. I.Linkov, F.K.Satterstrom, J.Steevens, E.Ferguson, R.C.Pleus, 'Multicriteria decision analysis and environmental risk assessment for nanomaterials', Journal of Nanoparticle Research 9: 543-554, 2007.

44. Igor Pujalté, Isabelle Passagn, Brigitte Brouillaud, Mona Trégue, Etienne Durand, Céline Ohayon-Courtès, and Béatrice L'Azou, 'Cytotoxicity and oxidative stress induced by different metallic nanoparticles on human kidney cells', Particle and Fibre Toxicology 8:10, 2011.

45. J.E.Can~as, B.Qi B, S.Li, J.D.Maul, S.B.Cox, S.Das, and M.J.Green, 'Acute and reproductive toxicity of nano-sized metal oxides (ZnO and TiO_2) to earthworms (*Eisenia fetida*)', Journal of Environmental Monitoring 13: 3351–3357, 2011.

46. J.Fabrega, S.N.Luoma, C.R.Tyler, T.S.Galloway, and J.R.Lead, 'Silver nanoparticles: behavior and effects in aquatic environment', Environment International 37: 517-31, 2011.

47. J.Hock, H.Hoffman, H.Krug, C.Lorenz, C.Limbach, B.Nowack, M.Riediker, K.Schischke, C.Som, W.Stark, C.Studer, N.von Gotz, S.Wengert, P.Wick, 'Guidelines on the precautionary Matrix for synthetic nanomaterials', Federal Office for Public Health and Federal Office for the Environment, 2008.

48. J.J.Scott-Fordsmand, and J.M.Weeks, 'Biomarkers in earthworms', Review of Environmental Contamination and Toxicology 165: 117–159, 2000.

49. J.Lead and E.Smith, 'Environmental and human health impacts of nanotechnology', Wiley, 2009.

50. J.M.Pettibone, A.Adamcakova-Dodd, P.S.Thorne, P.T.O'Shaughnessy, J.A.Weydert, and V.H.Grassian, 'Inflammatory response of mice following inhalation exposure to iron and copper nanoparticles', Nanotoxicology 2:189–204, 2008.

51. J.R.Gurr, S.S.Alexander, Wang, Chien-Hung Chen, and Kun-Yan Jan, 'Ultrafine titanium dioxide particles in the absence of photoactivation can induce oxidative damage to human bronchial epithelial cells', Toxicology 213: 66-73, 2005.

52. J.Venkateswara Rao and P.Kavitha, 'Toxicity of azodrin on the morphology and acetylcholinesterase activity of the earthworm (*Eisenia foetida*)', Environmental Research 96 (3): 323–327, 2004.

53. J.Y.Roh, Y.K.Park, K.Park, and J.Choi, 'Ecotoxicological investigation of CeO_2 and TiO_2 nanoparticles on the soil nematode *Caenorhabditis* using gene expression, growth, fertility and survival as endpoint', Environmental Toxicology and Pharmacology 29: 167–172, 2010.

54. Jana Petković, Bojana Žegura and Metka Filipič, 'Influence of TiO_2 nanoparticles on cellular antioxidant defense and its involvement in genotoxicity in HepG2 cells', Journal of Physics: Conference Series Volume 304 Number, 2011.

55. Janez Valant, Damjana Drobne and Sara Novak, 'Effect of ingested titanium dioxide nanoparticles on the digestive gland cell membrane of terrestrial isopods', Chemosphere, Volume 87, Issue 1, 19-25, 2012.

56. K.A.Guzman, M.R.Taylor, and M.F.Banfield, 'Environmental risks of nanotechnology: national nanotechnology initiative funding, 2000–2004', Environmental Science and Technology 40: 1401–1407, 2006.

57. K.Hund-Rinke and M.Simon, 'Ecotoxic effect of photocatalyticactive nanoparticles TiO_2 on algae and daphnids', Environmental Science and Pollution Research 13: 225-232. 2006.

58. K.Powers, M.Palazuelos, B.M.Moudgil, and S.M.Roberts, 'Characterization of the size, shape and state of dispersion of nanoparticles for toxicological studies', Nanotoxicology 44: 7821-6, 2007.

59. K.Powers, S.Brown, V.Krishna, S.Wasdo, B.Moudgil, and S.Roberts, 'Research strategies for safety evaluation of nanomaterials. Part VI. Characterization of nanoscale particles for toxicological evaluation', Toxicological Sciences 90: 296-303, 2006.

60. K.T.Kim, S.J.Klaine, J.Cho, S.H.Kim, and S.D.Kim, 'Oxidative stress responses of *Daphnia magna* exposed to TiO_2 nanoparticles according to size fraction', Science of the Total Environment 408: 2268–2272, 2010.

61. K.Tiede, M.Hassellöv, E.Breitbarth, Q.Chaudhry, and B.A.Boxall, 'Considerations for environmental fate and ecotoxicity testing to support environmental risk assessments for engineered nanoparticles', Journal of Chromatography A 1216: 503-509, 2009.

62. Karl Fent, Ecotoxicity of engineered nanoparticles, F.H. Frimmel, R. Niebner (eds.), Nanoparticles in the Water Cycle, Springer-Verlag Berlin Heidelberg, 2010.

63. Khara.D.Greieger, Anders Baun and Richard Owen, 'Redefining risk priorities for nanomaterials', Journal of Nanoparticle Research 12: 383-392, 2010.

64. Komkrit Suttiponparnit, Jingkun Jiang, Manoranjan Sahu, Sirikalaya Suvachittanont, Tawatchai Charinpanitkul and Pratim Biswas, 'Role of Surface Area, Primary Particle Size, and Crystal Phase on Titanium Dioxide Nanoparticle Dispersion Properties', Nanoscale Research Letters, 6:27, 2011.

65. Kong, Seog, Graham and Lee, 'Experimental considerations on the cytotoxicity of nanoparticles', Nanomedicine 6(5): 929–941, 2011.

66. L.C.Renwick, K.Donaldson, and A.Clouter, 'Impairment of alveolar macrophage phagocytosis by ultrafine particles', Toxicology and Applied Pharmacology, Vol. 172, No 2, pp. 119-127, 2001.

67. L.Canesia, R.Fabbria, G.Galloa, D.Vallottob, A.Marcominib, and G.Pojanab, 'Biomarkers in *Mytilus galloprovincialis* exposed to suspensions of selected nanoparticles Nano carbon black, C60 fullerene, Nano-TiO_2, Nano-SiO_2', Aquatic Toxicology 100: 168–177, 2010.

68. L.Hanssen, B.Walhout, and R.van Est, 'Ten lessons for a nanodialogue: the Dutch debate about nanotechnology thus far', Rathenau Institute, 2008.

69. L.Hao, Z.Wang, and B.Xing, 'Effect of sub-acute exposure to TiO_2 nanoparticles on oxidative stress and histopathological changes in Juvenile Carp (*Cyprinus carpio*)', Journal of Environmental Science (China) 21(10): 1459-66, 2009.

70. L.K.Limbach, P.Wich, P.Manser, R.N.Grass, A.Bruinink, and W.J.Stark, 'Exposure of engineered nanoparticles to human lung epithelial cells: influence of chemical composition and catalytic activity on oxidative stress', Environmental Science and Technology 41 (11): 4158–4163, 2007.

71. M.A.Kiser, P.Westerhoff, T.Benn, Y.Wang, J.Pérez-Rivera, and K.Hristovski, 'Titanium nanomaterial removal and release from wastewater treatment plants', Environmental Science and Technology 43: 6757-6763, 2009.

72. M.Schaedle and J.A.Bassham, 'Chloroplast glutathione reductase', Plant Physiology 59: 1011-1112, 1977.

73. Marina Belen Romanello, and Maria M. Fidalgo de Cortalezzi, 'An experimental study on the aggregation of TiO_2 nanoparticles under

environmentally relevant conditions', Water Research, Volume 47, Issue 12, 3887-3898, 2013.

74. Mary C Buford, Raymond F Hamilton Jr and Andrij Holian, 'A comparison of dispersing media for various engineered carbon nanoparticles', Particle and Fibre Toxicology 4:6, 2007.

75. Michala.E.Pettitt and Jamie.R.Lead. 'Minimum physicochemical characterization requirements for nanomaterial regulation', Environment International 52: 41-50, 2013.

76. N.C.Mueller and B.Nowack, 'Exposure modeling of engineered nanoparticles in the environment', Environmental Science and Technology 42: 4447-4453, 2008.

77. N.Laosiripojana, W.Sutthisripok and S.Assabumrungrat, 'Synthesis gas production from dry reforming of methane over CeO_2 doped Ni/Al_2O_3: influence of the doping ceria on the resistance toward carbon formation. Chemical Engineering Journal 112 (1–3): 13–22, 2005.

78. O.H.Lowry, N. J.Rosebrough, A.L.Farr and R. J.Randall, 'Protein measurement with the Folin-Phenol reagents', Journal of Biological Chemistry 193: 265-275, 1951.

79. OECD Guideline for testing of chemicals, no 207, Earthworm acute toxicity tests. Organization for Economic Cooperation and Development, Paris, France, 1984.

80. OECD Guideline for testing of chemicals, no 222, Earthworm reproduction test (*Eisenia fetida andrei*), Organization for Economic Cooperation and Development, Paris, France, 2004.

81. P.R.Lockman, J.M.Koziara, R.J.Mumper, and D.D.Allen, 'Nanoparticle surface charges alter blood-brain barrier integrity and permeability', Journal of Drug Targeting, 12: 635-641, 2004.

82. Paul. J.A. Borm, David Robbins, Stephan Haubold, Thomas Kuhlbusch, Heinz Fissan, Ken Donaldson, Roel Schins, Vicki

Stone, Wolfgang Kreyling, Jurgen Lademann, Jean Krutmann, David Warheit and Eva Oberdorster, 'The potential risks of nanomaterials: a review carried out for ECETOC', Particle and Fibre Toxicololgy 3:11, 2006.

83. Paula.S.Tourinho, A.M.Cornelis, Van Gestel, Stephen Lofts, Claus Svendsen, Amadeu.M.V.M.Soares and Susana Loureiro, 'Metal-based nanoparticles in soil: Fate, Behavior, and Effects on soil invertebrates', Environmental Toxicology and Chemistry, Vol. 31, No. 8, 1679–1692, 2012.

84. Qiuli Wua,Wei Wanga, Yinxia Li, Yiping Li, Boping Ye, Meng Tang, and Dayong Wang, 'Small sizes of TiO_2-NPs exhibit adverse effects at predicted environmental relevant concentrations on nematodes in a modified chronic toxicity assay system', Journal of Hazardous Materials 243: 161– 168, 2012.

85. R.D.Handy, T.B.Henry, T.M.Scown, B.D.Johnston, and C.R.Tyler, 'Manufactured nanoparticles: their uptake and effects on fish mechanistic analysis', Ecotoxicology 17: 396-409, 2008.

86. R.Doshi, W.Braida, C.Christodoulatos, M.Wazne, and G.O'Connor.G, 'Nano-aluminum: transport through sand columns and environmental effects on plants and soil communities', Environmental Research 106: 296–303, 2008.

87. R.F.Domingos, N.Tufenkji, and K.I.Wilkinson, 'Aggregation of titanium dioxide nanoparticles: role of a fulvic acid', Environmental Science and Technology 43: 1282-1286, 2009.

88. R.J.Aitken, M.Q.Chaudhry, A.B.A.Boxall, and M.Hull, 'Manufacture and use of nanomaterials: current status in the UK and global trends', Occupational Medicine 56:300-6, 2006.

89. R.Kaegi, A.Ulrich, B.Sinnet, R.Vonbank, A.Wichser, S.Zuleeg, H.Simmler, S.Brunner, H.Vonmont, M.Burkhardt, and M.Boller,

'Synthetic TiO$_2$ nanoparticle emission from exterior facades into the aquatic environment', Environmental Pollution 156: 233-239, 2008.

90. R.Kohen and A.Nyska, 'Oxidation of biological systems: oxidative stress phenomena, antioxidants, redox reactions, and methods for their quantification', Toxicologic Pathology 6: 620-650, 2002.

91. R.Landsiedel, L.Ma-Hock, A.Kroll, D.Hahn, J.Schnekenburger, K.Wiench, and W.Wohlleben, 'Testing Metal-Oxide Nanomaterials for Human Safety', Advanced Materials 22 (24): 2601–2627, 2010.

92. Ran Liu, Lihong Yin, Yuepu Pu, Geyu Liang, Juan Zhang, Yaoyao Su, Zhiping Xiao, and Bing Ye, 'Pulmonary toxicity induced by three forms of titanium dioxide nanoparticles via intra-tracheal instillation in rats', Progress in Natural Science, Volume 19, Issue 5, 573-579, 2009.

93. Ritesh K. Shuklaa,Vyom Sharmaa, Alok K. Pandeya, Shashi Singh, Sarwat Sultanac, and Alok Dhawana, 'ROS-mediated genotoxicity induced by titanium dioxide nanoparticles in human epidermal cells', Toxicology In Vitro 25: 231–241, 2011.

94. S.A.Reinecke, and A.J.Reinecke, 'The impact of organophosphate pesticides in orchards on earthworms in the Western Cape, South Africa', Ecotoxicology and Environmental Safety, 66 (2): 244–251, 2007.

95. S.B.Lovern, J.R.Strickler, and R.Klaper, 'Behavioural and Physiological Changes in *Daphnia magna* when Exposed to Nanoparticle Suspensions (Titanium Dioxide, Nano-C60 and C60HxC70Hx)', Environmental Science and Technology 41: 4465-4470, 2007.

96. S.F.Hansen, E.S.Michelson, A.Kamper, P.Borling, F.tuer-Lauridsen, and A.Baun, 'Categorization framework to aid exposure assessment of nanomaterials in consumer products', Ecotoxicology 17: 438–447, 2008.

97. S.Hall, T.Bradley, J.T.Moore, T.Kuykindall, and L.Minella, 'Acute and chronic toxicity of nano-scale TiO_2 particles to freshwater fish, cladocerans, and green algae, and effects of organic and inorganic substrate on TiO_2 toxicity', Nanotoxicology 3: 91-97, 2009.

98. S.J.Klaine, P.J.J.Alvarez, G.E.Batley, T.F.Fernandes, R.D.Handy, D.Y.Lyon, S.Mahendra, M.J.McLaughlin and J.R.Lead, 'Nanomaterials in the environment: behavior, fate, bioavailability, and effects', Environmental Toxicology and Chemistry 27: 1825-1851, 2008.

99. S.Marklund and G.Marklund, 'Involvement of the superoxide anion radical in the autoxidation of pyrogallol and a convenient assay for superoxide dismutase', European Journal of Biochemistry, 469-74, 1974.

100. S.P.Zhou, C.Q.Duan, F.U.Hui, Y.H.Chen, X.H.Wang, and Z.F.Yu, 'Toxicity assessment for chlorpyrifos-contaminated soil with three different earthworm test methods', Journal of Environmental Sciences 19 (7): 854–858, 2007.

101. Si Li and Weiling Sun, 'A comparative study on aggregation/sedimentation of TiO_2 nanoparticles in mono- and binary systems of fulvic acids and Fe (III)', Journal of Hazardous Materials 197: 70– 79, 2011.

102. Susanna Sforzini, Marta Boeria, Alessandro Dagninoa, Laura Oliveri, Claudia Bolognesi, and Aldo Viarengo, 'Genotoxicity assessment in *Eisenia andrei* coelomocytes: A study of the induction of DNA damage and micronuclei in earthworms exposed to B[a] P and TCDD-spiked soils', Mutation Research 746: 35– 41, 2012.

103. T.M.Sager, D.W.Porter, V.A.Robinson, W.G.Lindsley, D.E.Schwegler-Berry, and V.Castranova, 'Improved method to disperse nanoparticles for *in vitro* and *in vivo* investigation of toxicity', Nanotoxicology 1: 118–29, 2007.

104. V.K.Sharma, 'Agregation and toxicity of titanium dioxide nanoparticles in aquatic environment review', Journal of Environmental Science and Health, Part A 44: 1485-1495, 2009.

105. V.Stone, B.Nowack, A.Baun, N.van den Brink, and F.V.D.Kammer, 'Nanomaterials for environmental studies: classification, reference material issues, and strategies for physico-chemical characterization', Science of the Total Environment 408: 1745–1754, 2010.

106. Vicki.L.Colvin, 'The potential environmental impact of engineered nanomaterials', Nature Biotechnology, Vol 20 No.10, pp 1166-1170, 2003.

107. W.F.Vevers and A.N.Jha, 'Genotoxic and cytotoxic potential of titanium dioxide (TiO$_2$) nanoparticles on fish cells in vitro', Ecotoxicology 17: 410–420, 2008.

108. X.Liu, G.Chen, and C.Su, 'Effects of material properties on sedimentation and aggregation of titanium dioxide nanoparticles of anatase and rutile in the aqueous phase', Journal of Colloid and Interface Science 363: 84-91, 2011.

109. X.Zhu, Y.Chang, and Y.Chen, 'Toxicity and bioaccumulation of TiO$_2$ nanoparticle aggregates in *Daphnia magna*', Chemosphere 78: 209–215, 2010.

110. Yon Ju-Nam and Jamie R. Lead, 'Manufactured nanoparticles: An overview of their chemistry, interactions and potential environmental implications', Science of the Total Environment 400: 396-414, 2008.

111. Z.Abbas, C.Labbez, S.Nordholm, and E.Ahlberg, 'Size-Dependent Surface Charging of Nanoparticles', Journal of Physical Chemistry C 112: 5715–5723, 2008.

112. Zhaoxia Ji, Xue Jin, Saji George, Tian Xia, Huan meng, Xiang Wang, Elizabeth Suarez, Haiyuan Zhang, Eric MV Hoek, Hilary Godwin, Andre E. Nel, and Jeffrey I. Zink, 'Dispersion and St   ability   of

TiO$_2$ Nanoparticles in Cell Culture Media', Environmental Science and Technology 44(19): 7309-7314, 2010.

113. Zouhir E. Allounia, Paul J. Hølb, Miguel A. Cauquic, Nils R. Gjerdeta, and Mihaela R. Cimpana, 'Role of physicochemical characteristics in the uptake of TiO$_2$ nanoparticles by fibroblasts', Toxicology In vitro 26: 469-479, 2012.

List of publications in conference proceedings

1. B.Siva Prasad, **Ashwini Sri Hari**, P.Sankar Ganesh. 'Physicochemical Characterization and Ecotoxicological Evaluation of TiO_2 Nanoparticles in Earthworm *Eisenia foetida*', The Toxicologist, pg-215, 53rd Annual Meeting and ToxExpo of The Society of Toxicology, Phoenix, Arizona, March 2014

2. B.Siva Prasad, **Ashwini Sri Hari**, P.Sankar Ganesh. 'Should we say NO to NaNO? Preliminary study to corroborate occurrence of nanoparticles in treated wastewater samples', Proceedings of the National Conference on Technology, Policy and Community: Small Experiments in Sustainability, Hyderabad, India, March 2014

3. B.Siva Prasad, **Ashwini Sri Hari**, P.Sankar Ganesh. Development of environmental biomarkers to monitor endocrine disrupting chemicals in water and wastewater, National Conference on Sustainable Water Resources Planning, Management and Impact of Climate Change, Hyderabad, India, April 2013

PS 834 Physicochemical Characterization and Ecotoxicological Evaluation of TiO_2 Nanoparticles in Earthworm *Eisenia foetida*

B. Siva Prasad, S. Ashwini and P. Sankar Ganesh. *Biological Sciences, Birla Institute of Technology and Science Pilani, Hyderabad Campus, Hyderabad, India.*

Wide spread application and use of nanoparticles such as TiO_2 confers their release into various compartments of environment thereby raising concerns about their health impacts. It is essential to assess the ecotoxicity and implications of TiO_2 on the flora, fauna and ecosystem as a whole. Although there are studies reporting toxicity of TiO_2, but specific findings on its effect and underlining toxicity mechanisms in terrestrial organisms such as earthworms remain poorly understood and documented. Hence in this study, ecotoxicological evaluation of the rutile form (10-100nm) of TiO_2 nanoparticles on earthworm, *Eisenia foetida* was conducted as per the modified filter paper test of OECD-207 guidelines. Physicochemical characterization of TiO_2 nanoparticles was carried out using zeta sizer and scanning electron microscope. Earthworms were exposed to a series of concentrations (0.05, 0.1, 0.15, 0.2 and 0.25 mg/cm^2) of TiO_2 nanoparticles. Mortality of earthworms was determined after 48h exposure to TiO_2. Activity of the enzymes, catalase, super oxide dismutase and glutathione reductase was assessed in whole body homogenate of the worms surviving after exposure. Lipid peroxidation was measured to define the levels of oxidative stress during toxicity assessment. Perturbations caused by TiO_2 nanoparticles resulted in the oozing of coelomic fluid and pale coloration of earthworm body. Mortality rate, activity of antioxidant enzymes and lipid peroxidation were influenced by size and charge of TiO_2 nano particles rather than concentration. Results of the study showed that agglomeration of TiO_2 nanoparticles is responsible for the variation in their size and charge. The finding is first of its kind to establish ecotoxicity of rutile form of TiO_2 nanoparticles on earthworms. Similar studies will eventually help in developing strategies to predict ecotoxicity of TiO_2 in the soil environment.

NATIONAL CONFERENCE
ON

Technology, Policy and Community
SMALL
EXPERIMENTS IN
SUSTAINABILITY
14th - 15th March 2014

DEPARTMENT OF HUMANITIES
AND SOCIAL SCIENCES

BITS Pilani
Hyderabad Campus

Proceedings of the National Conference on Technology, Policy and Community:
Small Experiments in Sustainability, 14-15 March 2014

Should we say NO to NaNO?
Preliminary study to corroborate occurrence of nanoparticles in treated wastewater samples

B. Siva Prasad, Ashwini Sri Hari, P. Sankar Ganesh*

Department of Biological Sciences, BITS Pilani, Hyderabad Campus,
Jawahar Nagar, Shameerpet Mandal, RR District, Hyderabad 500 078, Andhra Pradesh, India

1. Introduction

Extensive production and use of nanotechnology based products; thereby their release into environment via sewage is currently occurring [1]. Concentrations of nanomaterials in wastewater and wastewater treatment plants will inevitably increase in the future. Wastewater is one of the important sources of nanomaterial discharge into the environment. Various forms of the source include treated effluent, biosolids, and plant-generated aerosols. Therefore, information on the occurrence of nanoparticles in wastewater will be an important contribution to understand their behavior and fate in aquatic ecosystem.

Titanium dioxide (TiO_2), Silver (Ag), and Fullerenes (C_{60}) are currently the most produced and hence found in a wide range of commercial products [2]. Recent studies on exposure modeling of nanomaterials indicated that predicted concentrations of nano-TiO_2 in wastewater effluents (0.7–16 µg/L) were higher than the predicted no-effect concentration level (1 µg/L) [2]. As nanomaterials have been used in various personal care products and environmental technologies, they may find their entry into various compartments of environment. Detection of such nanomaterials in wastewater can assist in understanding their fate in the environment while passing through a wastewater treatment plant. Therefore, the current preliminary study was designed to find the particle size distribution in wastewater collected from 340MLD. Sewage Treatment Plant located at Amberpet, Hyderabad.

2. Materials and methods

2.1 Collection and hotplate digestion of wastewater sample

Wastewater samples were collected from a conventional activated sludge wastewater treatment facility at Amberpet, Hyderabad. American Public Health Association (APHA)[3] guideline No. 1060 A was followed to collect wastewater sample. The samples were subjected for hot plate digestion as per modified guideline 3030 G reported by Westerhoff et al. (2011)[4, 5]. In the hot plate digestion method, sample was added to a 150 mL polytetrafluoroethylene (PTFE or Teflon) beaker along with 10 mL of hydrogen peroxide and 2 mL of nitric acid. The beakers were heated at 120°C for 4 hours to digest the organics. The beakers were removed from the hot plate and allowed to cool. When 0.1 to 0.5 mL of solution remained, the beakers were removed from the hot plate and allowed to cool. Then the beakers were rinsed three times with a solution of 2% nitric acid in nanopure milli q water into a 25 mL volumetric flask. The samples were filtered through 0.22 micron filter paper before analysis by dynamic light scattering.

---

*Corresponding author: (O): 40 66303547; (M): 9652345167; (E): sangan@hyderabad.bits-pilani.ac.in

2.2 Characterization of wastewater sample by dynamic light scattering

Several light scattering methods are among the ensemble approaches, and they can provide representative size distribution characterization of particles at relatively low concentrations (ca. 10^7-10^8 mL^{-1}) in a suspension. Dynamic light scattering (DLS), also called photon correlation spectroscopy, measure the laser light scattering by particles to calculate their hydrodynamic radius (r) distribution of particles in a suspension. DLS provides statistical representative data about the hydrodynamic size of nanomaterials. In situ, real-time monitoring of particle size distribution by DLS provides useful information regarding the aggregation process and, at the same time, gives quantitative measurement on the size of the particle clusters formed. Therefore, DLS was performed to ascertain the particle size distribution in the wastewater samples. All DLS measurements were performed with a Malvern Instrument Zetasizer Nano Series (Malvern Instruments, Westborough, MA, USA) equipped with a He-Ne laser ($\lambda = 633$ nm, max 5 mW) and operated at a scattering angle of 173°. In all measurements, 1 mL of sample was placed in a 10 mm × 10 mm quartz cuvette and size distribution was measured in the given sample. Samples were diluted in 1:8 dilutions in milli q water to attain the optimum polydispersity index during size distribution measurement in the water samples.

3. Results and discussion

Raw sewage entering the plant showed the particle z-average size 398 nm (range of the particle size was found to occur between 69 to 398 d.nm), while aeration tank sample showed particle z-average size 1067 nm (range of the particle size was found to occur between 77 to 892 d.nm) as represented in Figure 1.

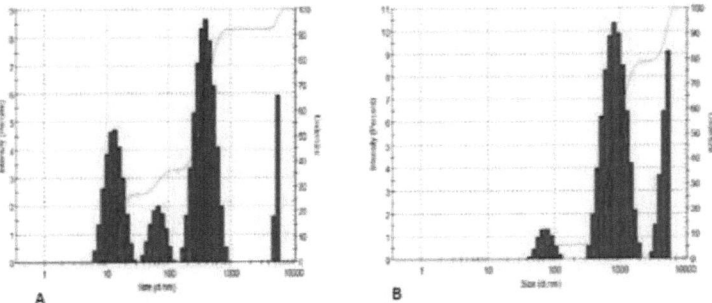

A

B

Figure 1: Size distribution of particles in wastewater sample of raw sewage (A) and aeration tank sample (B) of activated sludge determined by DLS. The Z-average of particle size distribution was calculated at particle distribution index (PdI) 0.37 and 0.54 for A and B samples respectively.

Samples showed the polydispersion due to their aggregation or sedimentation in the sample [6, 7]. Results suggest that significant fractions of nanoparticles are present in the wastewater. However, the interpretation of DLS data involves the interplay of other parameters, such as the size, concentration, shape, polydispersity, and surface properties of the particles involved. Hence, careful analysis of occurrence of nanoparticles in the sample should be confirmed by employing other confirmatory techniques such as nanoparticle tracking analysis, electron microscopy (SEM and TEM). It is essential to ensure the class or type of nanoparticle in samples by ICPMS techniques to conclude the species of materials present in the given sample [8-10].

4. Conclusions

This study can infer the basis of association of nanoparticles in wastewater samples. However, the relevant nanomaterials and their concentrations in wastewater samples can be established only by employing other analytical techniques such as ICPMS and electron microscopy. Thus, it is concluded that use of complementary techniques i.e. combination of mass spectrometry, light scattering and electron microscopy can provide substantial data to adjudge the characterization and quantification of nanomaterials in aquatic environment.

5. Acknowledgment

PSG and BSP thank Department of Biotechnology, Government of India for funding support in the form of Project No. BT/PR6592/GBD/27/446/2012 and Junior Research Fellowship, respectively. ASH thanks BITS Pilani, Hyderabad Campus for Teaching Assistantship.

6. References

1. Benn TM, Westerhoff PK. Nanoparticle silver released into water from commercially available sockfabrics. *Environ. Sci Technol* 2008, 42, 4133-4139.
2. Gottschalk. F.,Sonderer, T Scholz, R. W., Nowack, B. Modelled environmental concentrations of engineered nanomaterials (TiO₂, ZnO, Ag, CNT, Fullerenes) for different regions *Environ. Sci. Technol.* 2009, 43, 9216– 9222.
3. APHA, AWWA, WEF (1992). Standard methods for the examination of water and wastewater. 20ᵗʰ ed. USA.
4. Alex Weir, Paul Westerhoff, Lars Fabricius, Kiril Hristovski, and Natalie von Goetz. Titanium Dioxide Nanoparticles in Food and Personal Care Products *Environ. Sci Technol* 2012, 46, 2242-2250.
5. Paul Westerhoff, Guixue Song, Kiril Hristovski, Mehlika A. Kiser. Occurrence and removal of titanium at full scale wastewater treatment plants: Implications for TiO₂ nanomaterials. Journal of Environmental Monitoring 2011, 13, 1195-1203.
6. Donselaar LN. Philipse AP: Interactions between silica colloids with magnetite cores: diffusion sedimentation and light scattering. *J Colloid Interface Sci* 1999, 212, 14-23.
7. Phenrat T, Saleh N, Sirk K, Tilton RD, Lowry GV. Aggregation and sedimentation of aqueous nanoscale zerovalent iron dispersion. *Environ Sci Technol* 2007, 41, 284-290.
8. Nowack B, Bucheli T.D. Occurrence, behavior and effects of nanoparticles in the environment. Environ. Pollut. 2007, 150, 5-22.
9. Lombi, E., Donner, E., Tavakkoli, E., Turney, T.W., Naidu R., Miller B.W., Scheckel K.G. Fate of zinc oxide nanoparticles during anaerobic digestion of wastewater and post-treatment processing of sewage sludge. *Environ Sci Technol* 2012, 46(16), 9089-96.
10. Lorenz C., Windler L., von Goetz N., Lehmann R.P., Schuppler M., Hungerbühler K., Heuberger M., Nowack B. Characterization of silver release from commercially available functional (nano) textiles. *Chemosphere* 2012, 89(7), 817-24.

National Conference on

Sustainable Water Resources Planning, Management and Impact of Climate Change

April 05-06, 2013

Organized by

Centre of Excellence in Water Resources Management
Department of Civil Engineering

Birla Institute of Technology & Science, Pilani
Hyderabad Campus

Homepage: http://sites.bits-hyderabad.ac.in/swrm2013

DEVELOPMENT OF ENVIRONMENTAL BIOMARKERS TO MONITOR ENDOCRINE DISRUPTING CHEMICALS IN WATER AND WASTEWATER

B. Siva Prasad, Ashwini Sri Hari, P. Sankar Ganesh

Department of Biological Sciences, BITS Pilani, Hyderabad Campus

E-mail: sangan@hyderabad.bits-pilani.ac.in

In an era focused on sustainable development, safe supply of drinking water from sources of varying quality, including the reuse of wastewater is one of the key issues of water quality management. During the last few decades, the research that focused on water pollution has been largely directed towards the emerging contaminants, especially those displaying persistence in the environment. Endocrine disrupting chemicals (EDCs) are among the emerging contaminants that are widely distributed in the surface water have the potential to adversely affect the human health when the surface water is consumed along with water by the populace. The occurrence of the EDCs in surface water is becoming one of the major concerns worldwide. Continued release of these compounds into the environment even at low levels (ng L^{-1}) from wastewater treatment plants may show severe negative impacts on the function of ecosystems. EDCs are defined as exogenous substances that alter the function(s) of endocrine system and consequently cause adverse health effects in an intact organism, or its progeny, or sub - populations. EDCs can find their entry into aquatic environment by various routes including direct discharge from domestic, industrial and agricultural usage. Examples of EDCs include a wide variety of compounds ranging from pesticides (e.g. DDT, dieldrin and lindane), by-products of industrial processes (e.g. dioxins and PCBs), break down products of detergents (e.g. nonylphenol), compounds used in plastic manufacture (e.g. bisphenol A). Concerns about bioavailability and bioaccumulation of EDCs in drinking water and reclaimed wastewater have been considered as one of the key issues of many national and international health organizations. Recognizing EDCs as environmental health threat necessitates monitoring and evaluating the important public health issues raised by these chemicals in water and wastewater. Qualitative and quantitative measurements of EDCs in surface waters and wastewater effluents are essential to evaluate risks associated with their bioavailability in aquatic ecosystems. Though standard analytical methods are available for measuring variety of aquatic pollutants, they are primarily developed only for chemicals such as pesticides, metals, industrial chemicals, and polychlorinated biphenyls. However, the methods for analyzing endocrine activity are far less developed. This paper depicts the importance of developing biomarkers for monitoring EDCs in water and wastewater. Biomarkers approach will help in rapid monitoring of environmental impacts of EDCs.

Supplementary Data

Nanoparticle type	Test organism	Method of exposure	Exposure period	End points	Particle type	Nominal particle size (nm)	Concentrations used	TEM (nm)	SEM (nm)	DLS (nm)	SSA (BET) m²/g	Reference
TiO₂	*E.fetida*	Artificial soil test	7d	Bioaccumulation, Enzyme activity (SOD CAT & cellulase), MDA content, DNA	Rutile	10-20	0.1,0.5,1 & 5 g/Kg	NA	NA	NA	120-130	C.W.Hu *et al*,2010
TiO₂	*E.fetida*	Filter paper contact test, Sand acute test, Artificial soil test	14d	Acute and reproductive toxicity	Anatase	32	0.1, 1, 10, 100, 1000, 5000 and 10,000 (mg/L)	NA	Increase in aggregation with increase in conc. from 1-100mg / L		NA	E.Canas *et al*, 2011
Aged TiO₂ nanocomposite	*Lumbricus terrestris*	Water, ground horse manure & soil	7d	Bioaccumulation, Frequency of apoptosis	TiO₂ nanocom posite coated with Al(OH	100-300	1,10,100 mg/L (water); 10 & 100 mg/Kg (dry food); 15mg/ Kg (dry soil)	NA	NA	NA	NA	E.Lapied *et al*, 2011

Nanoparticle type	Test organism	Meth od of exposure	Exposure period	End points	Particle type	Nominal particle size (nm)	Concentra ions used	Characterization studies				Reference
								TEM (nm)	SEM (nm)	DLS (nm)	SSA (BET) (m²/g)	
TiO₂	E.fetida; E.andrei	Field test, Artifici al soil test	14d (motality), 18 weeks (juvenile growth), 28 d (reproducti on), and 48 h (avoidance)	Mortality, Reproduc tion, Avoidanc e, Juvenile growth	Anat ase, Anat ase + rutile	5 (Anat ase) 10 (Ana t ase) Anat ase+ R utile)	200-10,000 mg/Kg	20(+-7) Not disc erni ble 19(+-4)	N A	829 (p H 6. 8); 496 (p H 10) 805 1209	141 274 49	Mc.S h ane et al, 2012
TiO₂	E.fetida	Filter paper contact test	48h	Morta lity, Enzyme activity (CAT,G R & SOD), Lipid peroxida tion	Rutil e	10-100	0.05-0.25 mg/cm²	NA		Agglome ration of particles was seen	N A	B.Si va Prasa d et al, 201 4

Table. I Characterization studies carried out in previous TiO₂ nanoparticle ecotoxicity assessments in earthworms (Reference: Paula Tourinho *et al.*, 2012)

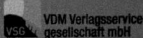